MW01484674

The Creative Curriculum® *for* Preschool

Teaching Guide

featuring the Buildings Study

Kai-leé Berke, Carol Aghayan, Cate Heroman

 **Teaching**Strategies® · Bethesda, MD

Copyright © 2010 by Teaching Strategies, LLC.

All rights reserved. No part of this text may be reproduced in any form or by any electronic or mechanical means, including information storage and retrieval systems, without prior written permission from Teaching Strategies, LLC, except in the case of brief quotations embodied in critical articles or reviews.

An exception is also made for the forms and the letters to families that are included in this guide. Permission is granted to duplicate those pages for use by the teachers/providers of the particular program that purchased these materials in order to implement *The Creative Curriculum® for Preschool* in the program. These materials may not be duplicated for training purposes without the express written permission of Teaching Strategies, LLC.

The publisher and the authors cannot be held responsible for injury, mishap, or damages incurred during the use of or because of the information in this book. The authors recommend appropriate and reasonable supervision at all times based on the age and capability of each child.

English editing: Lydia Paddock, Jayne Lytel
Design and layout: Jeff Cross, Amy Jackson, Abner Nieves
Spanish translation: Claudia Caicedo Núñez
Spanish editing: Judith F. Wohlberg, Alicia Fontán
Cover design: Laura Monger Design

Teaching Strategies, LLC
7101 Wisconsin Avenue, Suite 700
Bethesda, MD 20814

www.TeachingStrategies.com

978-1-60617-384-8

Library of Congress Cataloging-in-Publication Data

Berke, Kai-leé.
  The Creative Curriculum for Preschool Teaching Guide featuring the Buildings study / Kai-leé Berke, Carol Aghayan, Cate Heroman.
    p. cm.
  ISBN 978-1-60617-384-8
  1. Early childhood education--Curricula--United States. 2. Active learning--United States. 3. Buildings--Childhood and youth--United States.. 4. Structural engineering--Childhood and youth--United States. I. Aghayan, Carol. II. Heroman, Cate. III. Title.
  LB1139.4.B49 2010
  372.139 dc22

                    2010002938

Teaching Strategies and The Creative Curriculum names and logos are registered trademarks of Teaching Strategies, LLC, Bethesda, MD. This *Teaching Guide* is based on *The Creative Curriculum® Study Starters: Buildings* (Charlotte Stetson, lead author). Brand-name products of other companies are given for illustrative purposes only and are not required for implementation of the curriculum.

5 6 7 8 9 10 11 12   20 19 18 17 16 15 14
     Printing   Year Printed

Printed and bound in the United States of America

Acknowledgments

Many people helped with the creation of this *Teaching Guide* and the supporting teaching tools. We would like to thank Hilary Parrish Nelson for her guidance as our supportive Editorial Director and Jo Wilson for patiently keeping us on task. Both Hilary and Jan Greenberg provided a thoughtful and detailed content review that strengthened the final product.

Sherrie Rudick, Jan Greenberg, and Larry Bram deserve recognition for creating the first-ever children's book collection at Teaching Strategies, LLC. Working with Q2AMedia, they developed the concept for each book and saw the development process through from start to finish. Their hard work, creativity, patience, and attention to detail shines through in the finished product.

We are grateful to Dr. Lea McGee for her guidance, review, and feedback on our *Book Discussion Cards*. Jan Greenberg and Jessika Wellisch interpreted her research on a repeated read-aloud strategy to create a set of meaningful book discussion cards.

Thank you to Heather Baker, Toni Bickart, and Dr. Steve Sanders for writing more than 200 *Intentional Teaching Cards*, carefully aligning each teaching sequence with the related developmental progression and ensuring that children will receive the individualized instruction that they need to be successful learners. We are grateful to Sue Mistrett, who carefully reviewed each card and added strategies for including all children.

Translating *Mighty Minutes* into Spanish, ensuring cultural and linguistic appropriateness, was no easy task. Thank you to our dedicated team of writers and editors, including Spanish Educational Publishing, Dawn Terrill, Giuliana Rovedo, and Mary Conte.

Our brilliant editorial team, Toni Bickart, Lydia Paddock, Jayne Lytel, Diane Silver, Heather Schmitt, Heather Baker, Judy Wohlberg, Dawn Terrill, Giuliana Rovedo, Victory Productions, Elizabeth Tadlock, Reneé Fendrich, Kristyn Oldendorf, and Celine Tobal reviewed, refined, questioned, and sometimes rewrote our words, strengthening each page they touched.

Thank you to our Creative Services team for taking our words and putting them into a design that is both beautiful and easily accessible. The creative vision of Margot Ziperman, Abner Nieves, Jeff Cross, and Amy Jackson is deeply appreciated.

Our esteemed Latino Advisory Committee helped us continually reflect on how to support Spanish-speaking children and guided us through the development process. Thank you to Dr. Dina Castro, Dr. Linda Espinosa, Antonia Lopez, Dr. Lisa Lopez, and Dr. Patton Tabors.

We would like to acknowledge Lilian Katz and Sylvia Chard for their inspiring work on the Project Approach that has greatly advanced our thinking about quality curriculum for young children.

Most importantly, we would never be able to do this without the visionary guidance of Diane Trister Dodge. Her thoughtful leadership and commitment to young children and their families inspires all of the work that we do at Teaching Strategies.

Table of Contents

Getting Started

Why Investigate Buildings?

Young children are very curious about buildings. They want to know how they are constructed and what people do inside them. Perhaps you've seen a child mesmerized by large construction equipment as it moves massive piles of earth or hoists a steel beam high in the sky.

Buildings are everywhere in your community. They vary in size, color, construction, material, function, and location. Children enter and exit buildings every day as they leave and return home, spend time in other people's homes, go to school, and visit businesses. Sometimes children see buildings being constructed, repaired, or torn down.

This study offers many opportunities to explore buildings firsthand. Children will expand their knowledge and understanding of building materials and physical forces. They will also explore concepts in social studies related to shelter, jobs, and the purposes of different structures.

How do the children in your room show their interest in buildings? What do they say about buildings?

Web of Investigations

The *Teaching Guide Featuring the Buildings Study* includes five investigations aimed at exploring buildings. The investigations offer children an opportunity to learn more about the characteristics and features of buildings, the people who build them, and the role buildings play in our communities.

Some of the investigations also include site visits and visits to the classroom from guest speakers. Each investigation helps children explore science and social studies and strengthens their skills in literacy, math, technology, and the arts. Expand this web by adding your own ideas, particularly about aspects of the topic that are unique to your community.

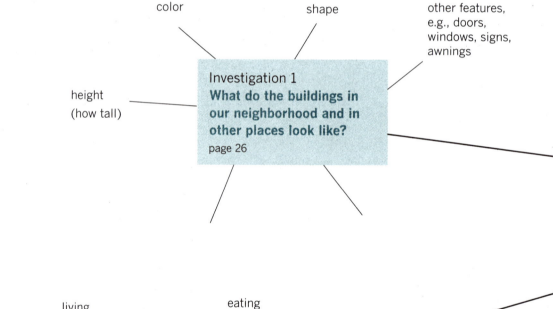

color

shape

other features, e.g., doors, windows, signs, awnings

height (how tall)

Investigation 1
What do the buildings in our neighborhood and in other places look like?
page 26

living

eating

Investigation 5
What happens inside buildings?
page 70

entertainment

community meetings

architects

carpenters

Investigation 2
**Who builds buildings?
What tools do they use?**
page 38

construction
vehicles

wood

glass

metal

Investigation 3
**What are buildings
made of? What makes
them strong?**
page 50

cement

design

hammer

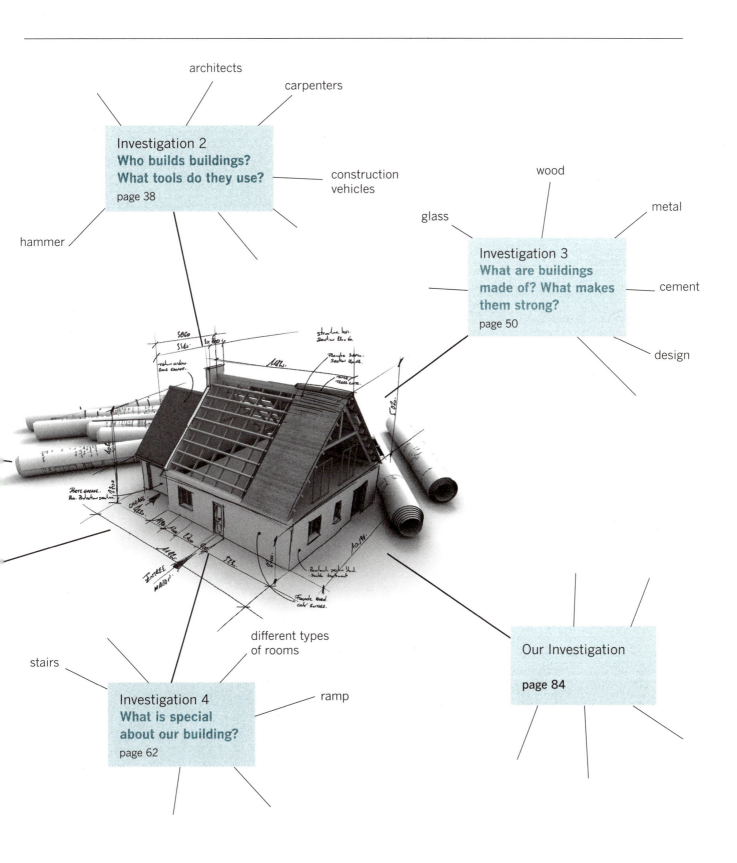

different types
of rooms

Our Investigation

page 84

stairs

Investigation 4
**What is special
about our building?**
page 62

ramp

A Letter to Families

Send families a letter introducing the study. Use the letter to communicate with families and as an opportunity to invite their participation in the study.

Dear Families,

When we are outside, children often ask questions about our building. They want to know the height of the school building, what it's made of, and how strong it is. They also ask many questions about the way it looks. Their ongoing interest in our building helped us realize that buildings would make a good study topic.

To get the study started, we are gathering all sorts of pictures of buildings. We could really use your help! We welcome pictures from any source, such as newspapers or magazines, postcards, printouts from the Internet, and your family's photo collection. It would be wonderful if you could include pictures of buildings in other parts of the world, too. Here's a list of suggestions, but you may also send in pictures of buildings that are not on the list.

houses	yurts	mosques	concert halls
apartments	shacks	post offices	museums
barns	hotels–motels	supermarkets	parking garages
sheds	inns	libraries	gas stations
castles	office buildings	restaurants	mechanic shops
cottages	schools	hospitals	government
cabins	stores	factories	buildings
bungalows	churches	skyscrapers	community centers
huts	synagogues	theaters	

As we study buildings, we will learn concepts and skills in science, social studies, literacy, math, the arts, and technology. We will also be using thinking skills to investigate, ask questions, solve problems, make predictions, and test our ideas.

What You Can Do at Home

Study your house or apartment building with your child. What materials were used to build it? How many floors or stories does it have? How many rooms, windows, and doors does it have? Don't forget the basement if you have one! How old is your home? Who built it? What are its dimensions?

Talk with your child about the buildings that you see together. Keep a list of the buildings that interest your child. Help your child investigate questions similar to the ones we mentioned above.

At the end of our study, we will have a special event to show you what we've learned. Thank you for playing an important role in our learning.

Carta a las familias

Envíe una carta a las familias para informarles sobre el estudio. Use la carta para comunicarse y como una oportunidad para invitarlas a participar.

Apreciadas familias,

Cuando estamos al aire libre, los niños a menudo hacen preguntas acerca de los edificios. Ellos desean saber la altura del edificio escolar, de qué está hecho y qué tan estable es. Además hacen preguntas acerca de cómo son. Por eso creemos que los edificios serían un buen tema de estudio.

Para poder comenzar el estudio estamos reuniendo toda clase de imágenes de edificios. Para hacer nuestra colección necesitamos de su ayuda. Si pueden colaborar, les agradeceríamos que nos envíen fotos, ilustraciones o imágenes de periódicos o revistas, postales, imágenes descargadas de internet y fotos de su familia. También sería maravilloso si pudiéramos incluir imágenes de edificios de otras partes del mundo. A continuación, les ofrecemos una lista de sugerencias, pero siéntanse libres de enviar imágenes de otros edificios que no estén en la lista. Las cuidaremos bien, para poder devolvérselas al final del estudio.

casas	hostales	oficinas del correo	auditorios
apartamentos	edificios de oficinas	supermercados	museos
graneros		bibliotecas	estacionamientos
cobertizos	escuelas	restaurantes	estaciones de gasolina
castillos	tiendas	hospitales	
cabañas	iglesias	fábricas	talleres de mecánica
chozas	sinagogas	rascacielos	edificios del gobierno
hoteles–moteles	mezquitas	teatros	centros comunitarios

Al estudiar los edificios aprenderemos conceptos y desarrollaremos destrezas en lecto-escritura, matemáticas, ciencia, estudios sociales, artes y tecnología. Además, desarrollaremos destrezas de razonamiento al investigar, hacer preguntas, resolver problemas, hacer predicciones y comprobar nuestras ideas.

Qué se puede hacer en el hogar

Hablen con los niños acerca de su casa o apartamento. ¿Qué materiales fueron usados para construirla? ¿Cuántos pisos o niveles tiene? ¿Cuántas habitaciones, ventanas y puertas tiene su vivienda? No olvide el sótano, si su casa tiene uno. ¿Cuántos años hace que fue construida? ¿Quién la construyó? ¿Qué dimensiones tiene?

Hablen con los niños acerca de los edificios que vean. Mantengan una lista de los edificios que les interesen a sus niños. Ayúdenles a investigar preguntas que tengan similares a las antes mencionadas.

Al finalizar nuestro estudio, tendremos un evento especial para mostrarles lo que aprendimos. De antemano agradecemos su participación y su importante rol en nuestro aprendizaje.

Beginning the Study

Introducing the Topic

To begin this study, you will explore the topic with the children to answer the following questions: What do we know about buildings? What do we want to find out about buildings?

Begin gathering pictures of many different types of buildings that you will use throughout the study. The pictures will show the outsides of buildings. Ask the children, their families, and friends to help you build the collection, which should include pictures of buildings from other parts of the world that look different from those in the children's community. A sample letter to families that includes information about sending in pictures appears in the beginning of this *Teaching Guide*. Below are some suggestions for gathering different kinds of building pictures.

Give each child a picture of a building. Take a walk around the outside of your school or program building and encourage children to think about the differences and similarities between the buildings in their pictures and their school. Later in the study, you can focus on the insides of buildings.

Find out whether the children have any family members in the building trades, such as construction workers, painters, architects, plumbers, electricians, or inspectors. If so, ask some of them to visit the classroom to talk to children about their work.

Allow several days for children to bring pictures of buildings to school. As children bring them in, display the pictures on a large wall. Record what children say about the pictures, and add their comments to the display.

> **What open-ended questions or prompts can you use to stimulate discussion about buildings with children?**

houses	yurts	mosques	concert halls
apartments	shacks	post offices	museums
barns	hotels–motels	supermarkets	parking garages
sheds	inns	libraries	gas stations
castles	office buildings	restaurants	mechanic shops
cottages	schools	hospitals	government
cabins	stores	factories	buildings
bungalows	churches	skyscrapers	community centers
huts	synagogues	theaters	

Preparing for Wow! Experiences

The "At a Glance" pages list these suggested Wow! Experiences, which require some advance planning.

Exploring the Topic:	Day 2: A walk around the outside of your building
Investigation 1:	Day 2: A walk around the neighborhood to look at different buildings
Investigation 2:	Day 2: A visit from someone who works in the construction field
	If possible, arrange for a visit to a construction site. Going on this visit will require some advance planning because of the safety precautions that must be taken. If there is a nearby construction project, ask the site supervisor whether the children may observe the work from across the street. Check to ensure that the children can sit in a safe place.
Investigation 3:	Day 2: A walk around the neighborhood to investigate the materials used to construct neighborhood buildings and to look for details that inspectors check
	Another idea is to arrange, if possible, a site visit to a very tall building. Try to interview the architect or engineer. If neither of these individuals is available, try to interview the building or maintenance manager.
Investigation 4:	Day 1: A walk around the inside of the school
	Day 2: An indepth investigation of an interesting feature of the school
	Day 3: A visit from a person who helps maintain the building
Investigation 5:	Day 2: A visit from a neighbor
	Day 3: A site visit to a different neighborhood building
Celebrating Learning:	Day 2: Building celebration

Exploring the Topic

What do we know about buildings? What do we want to find out?

	Day 1	Day 2	Day 3
Interest Areas	**Library:** books about building and construction **Blocks:** pictures of buildings **Computer:** eBook version of *The Three Little Pigs*	**Art:** pictures of buildings **Computer:** eBook version of *The Three Little Pigs*	**Toys and Games:** different kinds of open-ended connecting blocks
Question of the Day	Which building do you like best? (Display two different building photos.)	Which would you use to build your house: straw, sticks, or bricks?	How many doors are in our classroom?
Large Group	**Song:** "Scat Singing" **Discussion and Shared Writing:** Taking a Look at Buildings **Materials:** Mighty Minutes 14, "Scat Singing"; pictures of buildings	**Game:** Going on a Journey **Discussion and Shared Writing:** Exploring Our Building **Materials:** Mighty Minutes 63, "Going on a Journey"; your school building picture; Intentional Teaching Card LL45, "Observational Drawing"; small clipboards; black felt-tip pens	**Poem:** "A Building My Size" **Discussion and Shared Writing:** Parts of Buildings **Materials:** Mighty Minutes 49, "A Tree My Size"; small blocks or connecting blocks; "What do we know about buildings?" chart; pictures of buildings
Read-Aloud	*The Three Little Pigs*	*The Three Little Pigs* samples of straw, sticks, and brick	*Changes, Changes*
Small Group	**Option 1: More or Fewer Towers** Intentional Teaching Card M59, "More or Fewer Towers"; interlocking cubes; more–fewer spinner; numeral–quantity cards or die **Option 2: Which Has More?** Intentional Teaching Card M19, "Which Has More?"; ice cube trays or egg cartons; resealable sandwich bags; collection of objects that are similar in size	**Option 1: Dinnertime** Intentional Teaching Card M01, "Dinnertime"; paper or plastic dishes; utensils; napkins; cups; placemats **Option 2: Let's Go Fishing** Intentional Teaching Card M39, "Let's Go Fishing"; child-sized fishing poles; set of fish cards; paper clips	**Option 1: Counting & Comparing** Intentional Teaching Card M02, "Counting & Comparing"; pictures of buildings; card stock **Option 2: Counting & Comparing** Intentional Teaching Card M02, "Counting & Comparing"; pictures of buildings; digital camera; card stock
Mighty Minutes™	Mighty Minutes 47, "Step Up"; chart from large-group time	Mighty Minutes 57, "Find the Letter Sound"; letter cards	Mighty Minutes 3, "Purple Pants"; Mighty Minutes 08, "Clap the Missing Word"

Day 4	Day 5	Make Time For…
Blocks: pictures of buildings Computer: eBook version of *The Three Little Pigs*	Art: magazines with pictures of buildings; scissors Computer: eBook version of *The Three Little Pigs*	## Outdoor Experiences **Large Outdoor Blocks** • Bring some large building blocks or hollow blocks outdoors. • Invite children to use them to create buildings. • Provide empty cardboard boxes or crates if you don't have any large building blocks.
How many windows are in our classroom?	What do you want to know about buildings?	**Physical Fun** • Use Intentional Teaching Card P21, "Hopping." Follow the guidance on the card.
Movement: Skating **Discussion and Shared Writing:** What We Know About Buildings **Materials:** picture of roller or ice-skating rink; wax paper; "What do we know about buildings?" chart; pictures of buildings	Poem: "A Building My Size" **Discussion and Shared Writing:** What Do We Want to Find Out About Buildings? **Materials:** Mighty Minutes 49, "A Tree My Size"; "What do we want to find out about buildings?" chart; pictures of buildings	## Family Partnerships • Send home a letter to families introducing the study. • Invite families to contribute pictures of buildings in their neighborhoods. • Ask family members if they'd like to volunteer to accompany the class on the neighborhood walk during the first investigation.
The Three Little Pigs	*Keep Counting*	• Invite families to access the eBooks, *The Three Little Pigs* and *Keep Counting*.
Option 1: Alphabet Cards Intentional Teaching Card LL03, "Alphabet Cards"; alphabet cards; small manipulatives **Option 2: Buried Treasures** Intentional Teaching Card LL21, "Buried Treasures"; magnetic letters; large magnet; ruler or similar object; sand table with sand	**Option 1: Dramatic Story Retelling** Intentional Teaching Card LL06, "Dramatic Story Retelling"; *The Three Little Pigs*; props gathered during yesterday's read-aloud **Option 2: Clothesline Storytelling** Intentional Teaching Card LL33, "Clothesline Storytelling"; *The Three Little Pigs*; laminating supplies; 6 ft of clothesline; clothespins; a paper star; resealable bag	## Wow! Experiences • Day 2: A walk around the outside of your school building
Mighty Minutes 39, "Let's Pretend"	Mighty Minutes 15, "Say It, Show It"; numeral cards	

Day 1 Exploring the Topic

What do we know about buildings?
What do we want to find out?

Vocabulary

English: *collapse, buildings*

Spanish: *derrumbarse, edificios*

Large Group

Opening Routine

- Sing a welcome song and talk about who's here.

> See *Beginning the Year* for more information and ideas on planning your opening routine. See Intentional Teaching Card SE02, "Look Who's Here!" for attendance chart ideas.

Song: "Scat Singing"

- Use Mighty Minutes 14, "Scat Singing."

- Try the alliteration variation on the back of the card. Invite children to take the lead.

Discussion and Shared Writing: Taking a Look at Buildings

- Review the question of the day located on the "At a Glance" chart.

- Collect a variety of pictures of buildings.

- Pass around the collection of pictures. Say, "These are all pictures of *buildings*."

- Have each child select a picture.

- Invite children to talk about what they notice about the buildings in the pictures.

- Record their descriptions.

Before transitioning to interest areas, describe the books about building and construction in the Library area and talk about how children may use them.

Choice Time

As you interact with children in the interest areas, make time to

- With children in the Library area, look at books about building and construction. Pay attention to what they find interesting.

- Observe children's constructions in the Block area. Invite them to compare the structures they built with the pictures they selected today during large-group time.

- Listen for what children already know about buildings and what they're interested in learning more about. Record what they say and do.

Read-Aloud

Read *The Three Little Pigs.*

- **Before you read**, ask, "Who already knows something about this story?"

- **As you read**, introduce the word *collapse* as you talk about the pigs' houses.

- **After you read**, talk about the characters in the story, the problem the pigs faced, and the third little pig's solution to the problem. Tell the children that the book will be available on the computer in the Computer area.

Small Group

Option 1: More or Fewer Towers

- Review Intentional Teaching Card M59, "More or Fewer Towers." Follow the guidance on the card.

Option 2: Which Has More?

- Review Intentional Teaching Card M19, "Which Has More?" Follow the guidance on the card.

Children learn to compare amounts and use the word *more* very early in life. However, the words *fewer* and *less* are seldom a part of children's everyday vocabulary. Be sure to introduce these words intentionally as children engage in daily routines and class experiences.

Mighty Minutes™

- Use Mighty Minutes 47, "Step Up."

- Use the chart from large-group time. Try the letter sound variation on the back of the card. Offer support as needed.

Large-Group Roundup

- Recall the day's events.

- Invite children who looked at books about building and construction at choice time to share what they learned or enjoyed about the books.

What do we know about buildings?
What do we want to find out?

Vocabulary

English: *inspiration*
Spanish: *inspiración*

Large Group

Opening Routine

- Sing a welcome song and talk about who's here.

English-language learners
English-language learners must gain proficiency in both social and academic English to be successful in school. Social language enables children to interact effectively with others and form friendships. Academic language enables them to understand concepts and explain their thinking.

Game: Going on a Journey

- Review Mighty Minutes 63, "Going on a Journey." Follow the guidance on the card.

Discussion and Shared Writing: Exploring Our Building

- Show a picture of your school or center. Ask, "Who knows what building this is?"

- Record children's responses. Tell them that the picture shows their school.

- Explain, "We'll be walking around our building today to look at the outside."

- Point out something about the building that you are interested in seeing, e.g., say, "I'm interested in looking at the roof to see what it's made of."

- Ask, "What parts of our building are you interested in seeing?"

- Review Intentional Teaching Card LL45, "Observational Drawing." Follow the guidance on the card, and invite children to draw their observations of the building.

> While you're outside with the children exploring your building, give them enough time to draw it.

Before transitioning to interest areas, talk about the pictures of buildings in the Art area. Discuss how children can use the pictures as *inspiration* while painting at the easel.

Choice Time

As you interact with children in the interest areas, make time to

- Invite children to look through the pictures of buildings before they start painting. Remind them what *inspiration* means.

- Talk to children about their paintings. Ask questions about their choices and ideas.

Create a display of children's paintings along with the pictures of buildings that inspired them.

Read-Aloud

Read *The Three Little Pigs*.

- **Before you read**, recall the previous day's discussion about the characters, the problem, and the solution.

- **As you read**, review the question of the day.

- **After you read**, show samples of straw, sticks, and brick. Ask, "What do you notice about these three building materials?" Record children's responses.

English-language learners
Have pictures or samples of straw, sticks, and bricks available so that children who are not yet proficient in English can respond to the question of the day by pointing to the building materials.

Small Group

Option 1: Dinnertime

- Review Intentional Teaching Card M01, "Dinnertime." Follow the guidance on the card.

Option 2: Let's Go Fishing

- Review Intentional Teaching Card M39, "Let's Go Fishing." Follow the guidance on the card.

Mighty Minutes™

- Use Mighty Minutes 57, "Find the Letter Sound."

Large-Group Roundup

- Recall the day's events.
- Invite children who painted during choice time to share their work.
- Discuss today's walk around the school.

- Invite children to help you create a display of their observational drawings of the school.

Exploring the Topic

What do we know about buildings?
What do we want to find out?

Vocabulary

English: *wordless*

Spanish: *sin palabras*

Large Group

Opening Routine

- Sing a welcome song and talk about who's here.

Poem: "A Building My Size"

- Use Mighty Minutes 49, "A Tree My Size." Try the building variation on the back of the card.

Discussion and Shared Writing: Parts of Buildings

- Review the question of the day.

- Provide enough small blocks or connecting blocks for children to work in pairs.

- Invite children to work with a partner to construct a building. Display a few pictures of buildings so that children can refer to them for inspiration.

- After several minutes, ask, "Can you tell me about your building? What are the different parts of your building? How many windows does it have?"

- Record children's ideas on a chart titled "What do we know about buildings?"

English-language learners
Asking questions about the number of windows or doors in a building is a good way to teach number words in English and other languages spoken by the children. When a child responds by naming a number, invite the whole class to repeat that number in the language the child used.

Before transitioning to interest areas, talk about the connecting blocks in the Toys and Games area and how children may use them.

Choice Time

As you interact with children in the interest areas, make time to

- Talk to children about their buildings. Ask them about their choices and the process they used for making their decisions.

- Listen for what they know about buildings and record it on the "What do we know about buildings?" chart.

- Designate a safe place in the room where children can display their small building creations.

Read-Aloud

Read *Changes, Changes*.

- **Before you read**, tell the children the title of the book and show the cover. Ask, "What do you think this book is going to be about?"

- **As you read**, point out that this book is *wordless*—it doesn't have any words. Tell the story by describing what is happening in the pictures.

- **After you read**, recall children's predictions and discuss whether they were correct.

> Wordless books encourage children to use their imaginations to tell the story. Including wordless books in the Library area is ideal for family members who speak a language other than the language(s) in which most of the classroom book are written or whose limited reading abilities make reading aloud uncomfortable.

Small Group

Option 1: Counting & Comparing

- Review Intentional Teaching Card M02, "Counting & Comparing."

- Follow the guidance on the card using the collection of building pictures.

Option 2: Counting & Comparing

- Review Intentional Teaching Card M02, "Counting & Comparing."

- Go outside with the children and look at your building. Take pictures, including simple features, such as windows, doors, steps, and exterior lights.

- Follow the guidance on the card to count and compare your building's features.

Mighty Minutes™

- Use Mighty Minutes 03, "Purple Pants" and Mighty Minutes 08, "Clap the Missing Word." Follow the guidance on the cards.

Large-Group Roundup

- Recall the day's events.

- Invite children who constructed buildings during choice time to share their work.

Day 4   Exploring the Topic

What do we know about buildings?
What do we want to find out?

Vocabulary

English: *skate, elevator*

Spanish: *patinar, elevador*

Large Group

Opening Routine

- Sing a welcome song and talk about who's here.

Movement: "Skating"

- Show a picture of a roller- or ice-skating rink.

- Ask, "What do you think happens in this building?"

- Give children two pieces of waxed paper that are slightly bigger than their feet.

- Explain, "Let's pretend we are in this building. Let's also pretend that these pieces of waxed paper are our skates. Stand on your skates. Let's try to *skate* around the room. Keep your feet on your skates as you slide."

- Continue to invite children to skate.

- Ask children to throw their waxed paper away when they've finished using it.

> The waxed paper skates will work on any smooth type of flooring, such as carpet, wood, or laminate. The paper is somewhat slippery, so remind the children to use caution.

Discussion and Shared Writing:
What We Know About Buildings

- Review the "What do we know about buildings?" chart with the children.

- Pass around the collection of building pictures.

- Ask, "What else do we already know about buildings?"

- Record their ideas.

- Help children verbalize their ideas. Say, for example, "Gabriel, you said a tall building probably has an *elevator*. Do all buildings have *elevators*? So, some buildings have *elevators* and some don't. Let's write that on our chart."

- Review the question of the day.

- Invite children to think about other features of buildings that they already know about.

Before transitioning to interest areas, explain, "We have a big collection of building pictures. I need your help to organize them so we can hang them up in the Block area."

Choice Time

As you interact with children in the interest areas, make time to

- Sort the building pictures with children.

- Ask, "How can we sort these pictures? How are some of the buildings in them the same? How are they different?"

- Ask other questions that encourage children to figure out and explain their categories.

Read-Aloud

Read *The Three Little Pigs.*

- **Before you read**, say, "You already know a lot about this story."

- **As you read**, pause after reading a few pages and encourage children to retell the next couple of pages.

- **After you read**, ask, "If we were going to act out this story, what kinds of things would we need?" Record their ideas and help them gather the materials.

English-language learners
To encourage all children to participate in the discussion, have samples or illustrations of materials available so that children who are not yet proficient in English can point to their choices instead of naming them.

Small Group

Option 1: Alphabet Cards

- Review Intentional Teaching Card LL03, "Alphabet Cards." Follow the guidance on the card.

Option 2: Buried Treasures

- Review Intentional Teaching Card LL21, "Buried Treasures." Follow the guidance on the card.

Mighty Minutes™

- Use Mighty Minutes 39, "Let's Pretend." Follow the guidance on the card and encourage children to pretend that their bodies are elevators.

- Explain, "We are now at the first floor. Lower your body down to the ground so you'll be on the first floor. Now, someone press the number two. Let's move up to the second floor."

The words *first, second,* and *third,* etc., are ordinal numbers—numbers used to describe position in an ordered sequence. Most young children understand the concept of *first.* Other ordinal numbers are more difficult for them to learn.

Large-Group Roundup

- Recall the day's events.

- Invite children who sorted the pictures of buildings to describe the display in the Block area.

What do we know about buildings?
What do we want to find out?

Vocabulary

English: *investigate*

Spanish: *investigar*

Large Group

Opening Routine

- Sing a welcome song and talk about who's here.

Poem: "A Building My Size"

- Use Mighty Minutes 49, "A Tree My Size."

- Follow the guidance on the card using the building variation on the back.

Discussion and Shared Writing: What Do We Want to Find Out About Buildings?

- Say, "We already know a lot of things about buildings. Now let's think about what we want to find out about buildings. What questions would you like to *investigate*?"

- Review the question of the day. Add the children's responses to the "What do we want to find out about buildings?" chart.

- Model the questioning process for children. For example, show a picture of a big building and wonder aloud, "How could anyone build such a big building?" Your may also want to share a personal story about a time when you had a leaky roof, and say, "How do you think the roof got fixed?"

- Record any additional questions that the children would like to investigate.

- Help children formulate questions. For example, if a child says, "My house has a basement. I don't know how the house got on top of it," you might say, "Miguel, you're wondering how buildings are built on top of basements. I'll write, 'How are buildings with basements built?' on the chart."

Before transitioning to interest areas, describe the magazines in the Art area. Talk about how children may cut out pictures of buildings they find in them.

Choice Time

As you interact with children in the interest areas, make time to

- Look through the magazines with the children.

- Ask them questions and encourage them to describe the buildings they find in the pictures.

- Invite them to cut out pictures and add them to the display in the Block area.

- Offer to help tape some of the pictures to blocks for children to use in their constructions.

Use informal opportunities to observe a child's ability to use scissors and cut accurately. If children are having a hard time using scissors, give them opportunities to make their hands stronger. Examples of hand-strengthening activities include squeezing play dough, snipping paper of different thicknesses, participating in fingerplays that require opening and closing one's hands, and squeezing clothespins or sponges.

Read-Aloud

Read *Keep Counting*.

- **Before you read**, tell the children the title of the book and show the cover. Ask, "What do you think this book is about?"

- **As you read**, invite children to identify the numeral and count the buildings, vehicles, people, trees, etc. in the illustration on each page. Point out that as the numeral increases by one, the number of buildings, vehicles, and so on increases by one.

- **After you read**, ask children, "What else can you count up to 10 (fingers, toes, sneakers, crayons, etc.)?" Tell the children that the book will be available on the computer in the Computer area.

English-language learners
Count in children's home languages as well as in English. This helps English-language learners feel included and introduces new languages to other children.

Small Group

Option 1: Dramatic Story Retelling

- Review Intentional Teaching Card LL06, "Dramatic Story Retelling."

- Follow the guidance on the card to retell *The Three Little Pigs*. Use the materials you gathered during yesterday's read-aloud as props.

Option 2: Clothesline Storytelling

- Review Intentional Teaching Card LL33, "Clothesline Storytelling."

- Follow the guidance on the card to retell *The Three Little Pigs*.

Mighty Minutes™

- Use Mighty Minutes 15, "Say It, Show It."

Large-Group Roundup

- Recall the day's events.

- Invite children who cut out magazine pictures of buildings during choice time to share what they found.

Investigating the Topic

Introduction

You have already started lists of children's ideas and questions about buildings. As you implement the study, you will design investigations that help them expand their ideas, find answers to their questions, and learn important skills and concepts. This section has daily plans for investigating questions that children ask. Do not be limited by these suggestions. Use them as inspiration to design experiences tailored to your own group of children and the resources in your school and community. While it is important to respond to children's ideas and follow their lead as their thinking evolves, it is also important for you to organize the study and plan for possibilities. Be sure to review the "At a Glance" pages for suggested Wow! Experiences, as these events require some advance planning.

Investigation 1

What do the buildings in our neighborhood and in other

	Day 1	Day 2	Day 3
Interest Areas	Library: props and materials for retelling *The Three Little Pigs*	Toys and Games: different open-ended connecting blocks, e.g., ones that children have neither seen nor played with Computer: eBook version of *House, Sweet House*	Blocks: props related to buildings in your neighborhood, e.g., if you saw a fire station on the walk yesterday, add fire helmets; if you saw a farm, add farm animals
Question of the Day	Which of these buildings is near our school? (Display a picture of a building near the school and another building.)	How many buildings do you think we will see on our walk today?	What building did you like best? (Display a few pictures of neighborhood buildings seen on yesterday's walk.)
Large Group	**Song:** "My Body Jumps" **Discussion and Shared Writing:** Preparing for the Site Visit **Materials:** Mighty Minutes 72, "My Body Jumps"; teacher-created map of your neighborhood	**Song:** "Hi-Ho, the Derry-O" **Discussion and Shared Writing:** Making Building Predictions **Materials:** Mighty Minutes 23, "Hi-Ho, the Derry-O"; neighborhood map; Intentional Teaching Card SE01, "Site Visits"; digital camera; Intentional Teaching Card LL45, "Observational Drawing"; clipboards; paper; writing tools	**Movement:** Let's Stick Together **Discussion and Shared Writing:** Taking a Closer Look **Materials:** Mighty Minutes 67, "Let's Stick Together"; photos of neighborhood buildings; children's observational drawings
Read-Aloud	*A Chair for My Mother* Book Discussion Card 18 (first read-aloud)	*House, Sweet House*	*A Chair for My Mother* Book Discussion Card 18 (second read-aloud)
Small Group	**Option 1: We're Going on an Adventure** Intentional Teaching Card M36, "We're Going on an Adventure" **Option 2: Adventure Obstacle Course** Intentional Teaching Card M36, "We're Going on an Adventure"; items to build an obstacle course for the Block area	**Option 1: Story Problems** Intentional Teaching Card M22, "Story Problems"; collection of manipulatives to be added and subtracted **Option 2: Nursery Rhyme Count** Intentional Teaching Card M13, "Nursery Rhyme Count"; cotton balls or white pom-poms; green construction paper; numeral cards	**Option 1: Rhyming Riddles** Intentional Teaching Card LL11, "Rhyming Riddles"; props that rhyme with chosen words **Option 2: Rhyming Tubs** Intentional Teaching Card LL44, "Rhyming Tubs"; plastic tub; bag; pairs of objects with names that rhyme
Mighty Minutes™	Mighty Minutes 55, "Mr. Forgetful"	Mighty Minutes 18, "I'm Thinking Of…"	Mighty Minutes 38, "Spatial Patterns"

places look like?

Day 4	Day 5	Make Time For…
Library: books with pictures of buildings in other places **Computer:** eBook version of *Buildings, Buildings, Buildings*	**Art:** magazines; scissors	## Outdoor Experiences **Physical Fun** • Use Intentional Teaching Card P12, "Exploring Pathways." Follow the guidance on the card.
Have you ever seen a building like this? (Display a picture of an interesting building found in another country.)	We're making a pattern. Can you draw what comes next? (Draw the pattern door–window–door–window–door)	## Family Partnerships • Invite a family or community member who works on buildings to visit the classroom during Investigation 2, "Who builds buildings? What tools do they use?" This person could be an architect, electrician, engineer, a construction worker, painter, bricklayer, roofer, contractor, plumber, or anyone who constructs or maintains buildings. • Invite families to access the eBooks, *House, Sweet House* and *Buildings, Buildings, Buildings*
Movement: The Kids Go Marching In **Discussion and Shared Writing:** All Kinds of Homes **Materials:** Mighty Minutes 70, "The Kids Go Marching In"; *House, Sweet House*	**Poem:** "Two Plump Armadillos" **Discussion and Shared Writing:** Alike and Different **Materials:** Mighty Minutes 44, "Two Plump Armadillos"; pictures of buildings around the world; pictures of neighborhood buildings	## Wow! Experiences • Day 2: A walk around the neighborhood to look at different buildings
Buildings, Buildings, Buildings	*A Chair for My Mother* Book Discussion Card 18 (third read-aloud)	
Option 1: Letters, Letters, Letters Intentional Teaching Card LL07, "Letters, Letters, Letters"; alphabet rubber stamps; colored inkpads; construction paper **Option 2: Making My Name** Intentional Teaching Card LL29, "Making My Name"; small, sturdy envelopes; marker; letter manipulatives	**Option 1: Patterns** Intentional Teaching Card M14, "Patterns"; objects that arrange in a pattern; patterns examples; construction paper; crayons or markers **Option 2: Patterns on Buildings** Intentional Teaching Card M14, "Patterns"; objects that arrange in a pattern; patterns examples; construction paper; crayons or markers	
Mighty Minutes 12, "Ticky Ricky"; basket of items that children can name or identify	Mighty Minutes 25, "Freeze"; dance music	

Investigation 1

What do the buildings in our neighborhood and in other places look like?

Vocabulary

English: *map, neighborhood*

Spanish: *mapa, vecindario*

See Book Discussion Card 18, *A Chair for My Mother* (*Un sillón para mi mamá*), for additional words.

Large Group

Opening Routine

• Sing a welcome song and talk about who's here.

Song: "My Body Jumps"

• Use Mighty Minutes 72, "My Body Jumps."

Discussion and Shared Writing: Preparing for the Site Visit

• Say, "We've looked at our school building and lots of pictures of buildings. There are some interesting buildings here in our school's *neighborhood*."

• Talk about tomorrow's site visit to look at buildings near the school. Invite children to pay attention to the buildings they see as they walk around the neighborhood.

• Show children a simple, hand-drawn map of the neighborhood.

• Explain the map. Point out features, places, and important buildings. Say, for example, "Here is a *map* of our *neighborhood*. Here is our school, and here is the park at the end of our street." Save the map to use during Investigation 4, "What is special about our building?"

• Review the question of the day.

• Ask, "What other buildings are near our school?" Record their responses.

> The buildings you explore in your neighborhood don't have to be elaborate. Any building will offer something interesting for children to observe.

English-language learners
When introducing new words, such as *map* or *neighborhood*, determine whether the English words are a new way of labeling a concept that English-language learners already know. If not, teach children the concept (preferably in their home languages) as well as the new words.

Before transitioning to Interest areas, talk about the props and materials for retelling *The Three Little Pigs* in the Library area and how children may use them.

Choice Time

As you interact with children in the interest areas, make time to

- Observe children as they retell *The Three Little Pigs.*

- Notice children's ability to recall the characters, problem, and details of the story. Listen to the children's use of actual language from the text.

Read-Aloud

Read *A Chair for My Mother.*

- Use Book Discussion Card 18, *A Chair for My Mother.* Follow the guidance for the first read-aloud.

Small Group

Option 1: We're Going on an Adventure

- Review Intentional Teaching Card M36, "We're Going on an Adventure." Follow the guidance on the card.

Option 2: Adventure Obstacle Course

- Review Intentional Teaching Card M36, "We're Going on an Adventure."

- Build an obstacle course in the Block area to represent the actions in the song.

Obstacle courses are ideal for helping children develop gross-motor skills and learn positional words and phrases, such as *over, under, in between, behind, on top of,* and *next to.*

Mighty Minutes™

- Use Mighty Minutes 55, "Mr. Forgetful." Try the syllable variation on the back of the card if appropriate for your group of children.

Large-Group Roundup

- Recall the day's events.
- Invite children who retold *The Three Little Pigs* in the Library area to demonstrate how they used the props.

What do the buildings in our neighborhood and in other places look like?

Vocabulary

English: *map*

Spanish: *mapa*

Large Group

Opening Routine

- Sing a welcome song and talk about who's here.

Song: "Hi-Ho, the Derry-O"

- Use Mighty Minutes 23, "Hi-Ho, the Derry-O."

Discussion and Shared Writing: Making Building Predictions

- Review the walking route on the map you will follow through the neighborhood today.

- Ask, "What buildings do you think we will see? What do those buildings look like?"

- Record the children's responses.

- Review the question of the day and record children's responses.

- Review Intentional Teaching Card SE01, "Site Visits" and Intentional Teaching Card LL45, "Observational Drawing." Follow the guidance on both cards.

Before transitioning to interest areas, talk about the new connecting blocks, e.g., interlocking cubes or LEGO® bricks, in the Toys and Games area and how children may use them.

> If you don't have a *different* kind of connection block to introduce to the children, you may show a kind that has been available to the children for a while. Highlighting a familiar manipulative encourages children to think about new ways to use it.

Choice Time

As you interact with children in the interest areas, make time to

- Observe children as they construct their buildings in the Toys and Games area.

- Ask open-ended questions to encourage them to talk about their work.

- Photograph children's constructions or invite them to display the buildings in a protected area.

Read-Aloud

Read *House, Sweet House*.

- **Before you read**, say the title and show children the cover. Ask, "What do you think this book will be about?"

- **As you read**, invite children to comment on the pictures and story. Talk about the shapes of the houses.

- **After you read**, ask, "In what kind of building do you live?" Invite children to find a partner. Have the partners use building materials in the Block area, Art area, or Toys and Games area to construct one of the houses described in the book. Circulate among the interest areas. Talk about the shapes of the houses and the materials that children can use to construct them, e.g., say, "This house has a round dome on top, like part of a ball or sphere. Can you find something shaped like that in the classroom that you can use to build this house?" Tell the children that the book will be available on the computer in the Computer area.

Small Group

Option 1: Story Problems

- Review Intentional Teaching Card M22, "Story Problems." Follow the guidance on the card.

Option 2: Nursery Rhyme Count

- Review Intentional Teaching Card M13, "Nursery Rhyme Count." Follow the guidance on the card.

English-language learners
When calling on individual children in a small group, be sure to address them by name so they'll know that they are being asked to participate. This strategy will help them feel included.

Mighty Minutes™

- Use Mighty Minutes 18, "I'm Thinking Of..." Use items seen on the site visit.

Large-Group Roundup

- Recall the day's events.

- Invite children who constructed buildings in the Toys and Games area to share their work.

- Write a group thank-you note to the visitor who works in construction. Invite children to sign their names and add drawings to the note.

What do the buildings in our neighborhood and in other places look like?

Vocabulary

See Book Discussion Card 18, *A Chair for My Mother* (*Un sillón para mi mamá*) for words.

Large Group

Opening Routine

- Sing a welcome song and talk about who's here.

Movement: Let's Stick Together

- Use Mighty Minutes 67, "Let's Stick Together."

Discussion and Shared Writing: Taking a Closer Look

- Review the question of the day.

- Show photos of buildings from yesterday's walk. Also show the children's observational drawings.

- Ask questions to help children extend their thinking and discuss a building's details: "What do you think this building is made of? What do the doors and windows look like? What shape is the sign above the door? How is this brick building different from that one?"

- Record children's responses.

English-language learners
Use gestures and model language to help children answer these questions. You can also provide samples or illustrations of different shapes, materials, and objects. That way, children can respond by pointing.

Before transitioning to interest areas, talk about the props in the Block area. Remind children about the buildings they saw on yesterday's walk and talk about how the props relate to them. Show the props and ask, "What can you do with these?"

Choice Time

As you interact with children in the interest areas, make time to

- Observe children as they use the props in the Block area.

- Invite children to use the block building materials to construct buildings related to the props.

Read-Aloud	Read *A Chair for My Mother.*	
	• Use Book Discussion Card 18, *A Chair for My Mother.* Follow the guidance for the second read-aloud.	

Small Group	**Option 1: Rhyming Riddles**	**Option 2: Rhyming Tubs**
	• Review Intentional Teaching Card LL11, "Rhyming Riddles." Follow the guidance on the card.	• Review Intentional Teaching Card LL44, "Rhyming Tubs." Follow the guidance on the card.

Mighty Minutes™	• Use Mighty Minutes 38, "Spatial Patterns." Try the variation on the back of the card.

Large-Group Roundup	• Recall the day's events. • Invite children to recall some of the rhyming word pairs from small-group time.	• Record the words and then read them back so that the group can say them with you.

What do the buildings in our neighborhood and in other places look like?

Vocabulary

English: *similarities, differences*

Spanish: *semejanzas, diferencias*

Large Group

Opening Routine

- Sing a welcome song and talk about who's here.

Movement: The Kids Go Marching In

- Use Mighty Minutes 70, "The Kids Go Marching In."

Discussion and Shared Writing: All Kinds of Homes

- Reread *House, Sweet House.*

- Ask children to describe some of the buildings in the book.

- Discuss the characteristics of the homes.

- Invite children to describe similarities and differences. For example, say, "Houses have *similarities*. They are all places for families to live. Houses have *differences*, too. Some have one family living in them and some have more than one family in them."

- Record children's ideas.

- Using the information provided in the book, talk about why the buildings look the way they do.

> Be sensitive to children in your class who may be homeless. Bear in mind that you will not necessarily know which families in your program are in that situation. Consider also the children's different socioeconomic backgrounds and what those differences mean in terms of the kinds of homes they live in. Be careful to focus this discussion on the idea that homes are places for families to live together, no matter what their homes look like.

Before transitioning to interest areas, review the question of the day. Talk about the books in the Library area. Explain, "These books have pictures of buildings that are not in our community." Show some of the pictures and talk about their interesting characteristics.

Choice Time

As you interact with children in the interest areas, make time to

- Look through the books about buildings with children in the Library area.

- Read aloud to them any information about buildings that captures their interest.

Read-Aloud

Read *Buildings, Buildings, Buildings.*

- **Before you read**, show the cover of the book and invite children to guess what happens inside the buildings displayed on the cover.

- **As you read**, pause to invite children to talk about buildings that are familiar to them.

- **After you read**, check children's predictions about the building on the cover. Tell the children that the book will be available on the computer in the Computer area.

Small Group

Option 1: Letters, Letters, Letters

- Review Intentional Teaching Card LL07, "Letters, Letters, Letters." Follow the guidance on the card.

Option 2: Making My Name

- Review Intentional Teaching Card LL29, "Making My Name." Follow the guidance on the card.

Mighty Minutes™

- Use Mighty Minutes 12, "Ticky Ricky." Try the pattern variation on the back of the card.

Large-Group Roundup

- Recall the day's events.

- Invite children who looked at books in the Library area to describe for the group any interesting buildings they discovered.

What do the buildings in our neighborhood and in other places look like?

Vocabulary

English: *alike, different*

Spanish: *parecidos, diferentes*

See Book Discussion Card 18, *A Chair for My Mother* (*Un sillón para mi mamá*), for additional words.

Large Group

Opening Routine

- Sing a welcome song and talk about who's here.

Poem: "Two Plump Armadillos"

- Use Mighty Minutes 44, "Two Plump Armadillos."

Discussion and Shared Writing: Alike and Different

- Remind children about some of the homes in *House, Sweet House* that were built in other places around the world. Show several pictures of buildings in the neighborhood.

- Invite children to compare the pictures of neighborhood buildings with the pictures of buildings from around the world. Ask, "How are these buildings *alike*? How are they *different*?"

- Record children's responses.

Before transitioning to interest areas, talk about the magazines in the Art area. Explain that children may use them to find and cut out pictures of interesting buildings.

Choice Time

As you interact with children in the interest areas, make time to

- Talk to the children about the buildings they find in the magazines. Pay attention to what they find interesting.

- Read aloud any information the magazines provide about those buildings.

- Help children add the pictures they've cut out to the display in the Block area.

Read-Aloud

Read *A Chair for My Mother.*

- Use Book Discussion Card 18, *A Chair for My Mother.* Follow the guidance for the third read-aloud.

Small Group

Option 1: Patterns

- Talk about the question of the day.
- Review Intentional Teaching Card M14, "Patterns." Follow the guidance on the card.

Option 2: Patterns on Buildings

- Talk about the question of the day.
- Review Intentional Teaching Card M14, "Patterns." Follow the guidance on the card.
- Invite children to explore the pictures of buildings and look for patterns on the buildings.

English-language learners
To help children who are English-language learners participate in the activity, explain your actions as you make them and explain what other children are doing. For example, say, "I am drawing a door. I am drawing a window. Now I am drawing a door again. I'm drawing a window again." Next, pointing to each object as you name it, say, "I am making a pattern: door, window, door, window."

Mighty Minutes™

- Use Mighty Minutes 25, "Freeze." Try the letter or word variation on the back of the card.

Large-Group Roundup

- Recall the day's events.
- Invite children who added pictures to the Block area display to point out the buildings and describe them to the group.

Investigation 2

Who builds buildings? What tools do they use?

	Day 1	Day 2	Day 3
Interest Areas	**Blocks:** measuring tapes; carpenter's rulers; T-squares **Computer:** eBook version of *Build It From A to Z*	**Blocks:** tool belt; hard hat; toolkit	**Art:** hammer; nails; soft wood, such as pine **Computer:** eBook version of *Build It From A to Z*
Question of the Day	Can you build a building as tall as you?	What would you like to ask our visitor today?	What can you do with a hammer?
Large Group	**Game:** What's Inside the Box? **Discussion and Shared Writing:** Who Builds Buildings? **Materials:** Mighty Minutes 31, "What's Inside the Box?"; tape measure; box; pictures of a construction site	**Song:** "This Is the Way" **Discussion and Shared Writing:** Expert Interview **Materials:** Mighty Minutes 06, "This Is the Way"	**Song:** "The People in Your Neighborhood" **Discussion and Shared Writing:** Tools **Materials:** Mighty Minutes 01, "The People in Your Neighborhood"; bag; tools
Read-Aloud	*Build It From A to Z* chart from large-group time	*Building a House* chart from day 1 large-group time and read-aloud	*Build It From A to Z* chart from day 1 large-group time
Small Group	**Option 1: Bookmaking** Intentional Teaching Card LL04, "Bookmaking"; *A Chair for My Mother*; clipboards; cardboard or card stock; blank paper; pencils, crayons, or markers; bookbinding supplies **Option 2: Desktop Publishing** Intentional Teaching Card LL02, "Desktop Publishing"; *A Chair for My Mother*; digital camera; computer; each child's word bank; printer; paper; bookbinding supplies	**Option 1: Bookmaking** Intentional Teaching Card LL04, "Bookmaking"; *A Chair for My Mother*; clipboards; cardboard or card stock; blank paper; pencils, crayons, or markers; bookbinding supplies **Option 2: Desktop Publishing** Intentional Teaching Card LL02, "Desktop Publishing"; *A Chair for My Mother*; digital camera; computer; each child's word bank; printer; paper; bookbinding supplies	**Option 1: Alphabet Books** Intentional Teaching Card LL34, "Alphabet Books"; *Build It From A to Z*; construction paper; alphabet cards **Option 2: Alphabet Books and Tools** Intentional Teaching Card LL34, "Alphabet Books"; *Build It From A to Z*; tools; construction paper; alphabet cards
Mighty Minutes™	Mighty Minutes 04, "Riddle Dee Dee"	Mighty Minutes 97, "Shape Hunt"; three-dimensional shapes or shape cards	Mighty Minutes 22, "Hot or Cold 3-D Shapes"; several three-dimensional shapes

Day 4	Day 5	Make Time For…
Blocks: blueprints; building floor plans; clipboards; paper and pencils; digital camera	Sand and Water: small construction vehicles Intentional Teaching Card SE14, "Playing Together"	## Outdoor Experiences **Observing a Building's Shadow** • On a sunny day, go outside and point out children's shadows.
When you grow up, what job would you like to try? (Display photos of some of the jobs from the list on day 1.)	What part of your body does this protect? (Display a hard hat.)	• Talk to children about their shadows. • Point out that the school building also casts a shadow.
Song: "Pound the Nail" **Discussion and Shared Writing:** Building Jobs **Materials:** Mighty Minutes 43, "Bouncing Big Brown Balls"; list of people who work on buildings from the day 1 large-group discussion	Song: "What Is My Job?" **Discussion and Shared Writing:** Builders Staying Safe **Materials:** Mighty Minutes 11, "What Is My Job?"; *Build It From A to Z*; hard hat	• Invite children to observe the school's shadow and trace it with sidewalk chalk. Talk about the size of the shadow compared to the size of the school building. **Physical Fun** • Use Intentional Teaching Card P11, "Jump the River." Follow the guidance on the card.
Changes, Changes	*The Three Little Javelinas* Book Discussion Card 19 (first read-aloud)	## Family Partnerships • Invite families to accompany the class on a neighborhood walk during Investigation 3, "What are buildings made of? What makes buildings strong?"
Option 1: Show Me Five Intentional Teaching Card M16, "Show Me Five"; small building-related manipulatives, such as nuts, bolts, and wood scraps **Option 2: Guessing Jar** Intentional Teaching Card M17, "Guessing Jar"; large plastic jar; small building-related manipulatives, such as nuts, bolts, and wood scraps	**Option 1: Cube Trains** Intentional Teaching Card M40, "Cube Trains"; interlocking cubes; pictures or illustrations of trains **Option 2: Action Patterns** Intentional Teaching Card M35, "Action Patterns"; action cards; pocket chart	• Invite families to access the eBook, *Build It From A to Z*. ## Wow! Experiences • Day 2: Visit from someone who works in the construction field
Mighty Minutes 60, "The Name Dance"	Mighty Minutes 59, "Clap the Beat"; building-related pictures or objects	

Who builds buildings?
What tools do they use?

Vocabulary

English: *construction site, tape measure*

Spanish: *un lugar en construcción, cinta métrica*

Large Group

Opening Routine

• Sing a welcome song and talk about who's here.

Game: What's Inside the Box?

• Use Mighty Minutes 31, "What's Inside the Box?"

• Follow the guidance on the card, using a tape measure as the object in the box.

Discussion and Shared Writing: Who Builds Buildings?

• Show pictures of a construction site.

• Invite the children to talk about what they notice in the pictures.

• Ask, "Who builds buildings?"

• Record their responses.

• Save this chart to use during the read-aloud.

Before transitioning to interest areas, talk about the tape measure and other measuring tools in the Block area and how children may use them.

Choice Time

As you interact with children in the interest areas, make time to

• Discuss the question of the day with the children in the Block area.

• Explain how they can use the measuring tools to compare their heights with the heights of their buildings.

Read-Aloud

Read *Build It From A to Z.*

- **Before you read**, take a "picture walk" through the book. This strategy involves talking about the pictures in the story and asking the children to predict what they will learn from the book.

- **As you read**, point out the different people in the book who work on buildings.

- **After you read**, add those people to the list that you started during large-group time. Save this list for the day 4 large-group discussion. Tell the children that the book will be available on the computer in the Computer area.

> **For more strategies on reading aloud to groups of children, see *The Creative Curriculum® for Preschool, Volume 3: Literacy*.**

Small Group

Option 1: Bookmaking

- Remind the children how the little girl in *A Chair for My Mother* really wanted her mother to have a comfortable chair.

- Ask, "Have you ever wanted to get something for someone else? What did you want to get? To whom did you want to give it?"

- Use Intentional Teaching Card LL04, "Bookmaking," to help children make a book that reflects their answers.

Option 2: Desktop Publishing

- Remind the children how the little girl in *A Chair for My Mother* really wanted her mother to have a comfortable chair.

- Ask, "Have you ever wanted to get something for someone else? What did you want to get? To whom did you want to give it?"

- Use Intentional Teaching Card LL02, "Desktop Publishing," to help children make a book that reflects their answers.

Mighty Minutes™

- Use Mighty Minutes 4, "Riddle Dee Dee." Try the jumping numbers variation on the back of the card.

Large-Group Roundup

- Recall the day's events.
- Invite children who constructed buildings in the Block area to share their discoveries.
- Explain, "Someone who works on buildings will be coming to our classroom tomorrow so we can interview him. Let's think about some questions we'd like to ask him."
- Record children's questions.

English-language learners
When English-language learners try to participate by saying a single word, respond by integrating that word into a few sentences. For example, if a child says, "Tool," ask, "Do you want to know what tools our visitor uses in her job? Asking about tools is an important question." Expanding on the child's language helps him or her learn more about English vocabulary and structure.

Who builds buildings?
What tools do they use?

Vocabulary

English: *pound, twist, saw*

Spanish: *clavar, dar vuelta, serruchar*

Large Group

Opening Routine

• Sing a welcome song and talk about who's here.

Song: "This Is the Way"

• Use Mighty Minutes 06, "This Is the Way."

• Follow the guidance on the card using carpentry-related phrases, e.g., saw the wood, pound the nail, paint the wall, twist the screw, and lay the brick.

Discussion and Shared Writing: Expert Interview

• Review the question of the day. Add any new questions to the chart from yesterday's large-group roundup.

• Introduce the visitor.

• Ask him to explain his job and show some of the tools that he uses.

• Encourage children to ask the questions they generated in large-group roundup yesterday.

• Record the visitor's responses.

English-language learners
If the visitor speaks any of the children's home languages, ask him to respond to children's questions in English and in any other language(s) in which he is proficient. Supporting children's home-language development helps them keep their cultural identity, stay attached to family traditions, and become bilingual.

Before transitioning to interest areas, talk about the construction props in the Block area and how children may use them.

> Adding a few props to the Block area stimulates a child's imagination and expands her dramatic play about new topics. Take care not to add too many new props at one time as this can be overwhelming and actually make the block-building experience less engaging and meaningful for children.

Choice Time

As you interact with children in the interest areas, make time to

- Observe children as they play with the props in the Block area.

- Ask children questions to extend their play, e.g., "Who works in your building? What do the workers do there?"

> **Transition children to interest areas using Mighty Minutes 47, "Step Up." Have them circle a word on the chart and tell you which interest area they are going to.**

Read-Aloud

Read *Building a House*.

- **Before you read**, show the cover of the book and read the title. Explain, "This book shows us the steps for building a house." Ask, "What do you think workers need to do to build a house?"

- **As you read**, point out the steps mentioned in the book that match children's predictions.

- **After you read**, review all of the workers mentioned in the book. Add to the list from yesterday if some are not already included.

Small Group

Option 1: Bookmaking

- Remind the children how the little girl in *A Chair for My Mother* really wanted her mother to have a comfortable chair.

- Ask, "Have you ever wanted to get something for someone else? What did you want to get? To whom did you want to give it?"

- Use Intentional Teaching Card LL04, "Bookmaking," to help children make a book that reflects their answers.

Option 2: Desktop Publishing

- Remind the children about how the little girl in *A Chair for My Mother* really wanted her mother to have a comfortable chair.

- Ask, "Have you ever wanted to get something for someone else? What did you want to get? To whom did you want to give it?"

- Use Intentional Teaching Card LL02, "Desktop Publishing," to help children make a book that reflects their answers.

Mighty Minutes™

- Use Mighty Minutes 97, "Shape Hunt." Follow the guidance on the card.

Large-Group Roundup

- Recall the day's events.

- Read the books that the children created over the last two days during small-group time.

Day 3   Investigation 2

Who builds buildings?
What tools do they use?

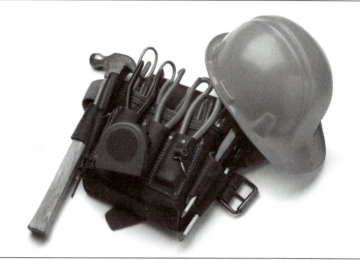

Vocabulary

English: *blueprint*

Spanish: *plano*

Large Group

Opening Routine

• Sing a welcome song and talk about who's here.

Song: "The People in Your Neighborhood"

• Use Mighty Minutes 01, "The People in Your Neighborhood."

• Follow the guidance on the card using construction workers, e.g., carpenter, bricklayer, or plumber.

Discussion and Shared Writing: Tools

• Place a hammer inside a mystery bag.

• Invite children to reach inside the bag and describe what they feel.

• Record their descriptions.

• Remove the item and name it. Say, "Michael said it was hard. Javier said it had a long part and a short part. Rosa said it was cold. Let's see what it is. It's a hammer!" Record the tool's name above the children's descriptions.

• Talk about how the tool is used.

• Review the question of the day.

• Repeat this activity with other tools.

English-language learners
When recording information about the tools, include both pictures and words. Write the words—especially the name of the tool—in English and in children's home languages, if you are able.

Before transitioning to interest areas, talk about the hammer, nails, and wood in the Art area and how children may use them.

Before allowing children to use woodworking tools, make sure they know and can demonstrate safe handling procedures. Consider asking a parent or other adult to supervise an activity. Review the suggested safety rules described in the Art area chapter in *The Creative Curriculum® for Preschool, Volume 2: Interest Areas.*

Choice Time

As you interact with children in the interest areas, make time to

- Observe children as they work with the wood. Ask questions about their constructions.

- Offer assistance and safety reminders as needed.

> For ideas on enforcing safety rules, see Intentional Teaching Card SE09, "Big Rule, Little Rule."

Read-Aloud

Read *Build It From A to Z.*

- **Before you read**, ask, "What do you remember about this book?"

- **As you read**, point out the blueprints and talk about how they are used. Add *blueprints* to the list of tools from large-group time.

- **After you read**, review the tools in the book. Add to the list of tools from large-group time as needed.

Small Group

Option 1: Alphabet Books

- Review Intentional Teaching Card LL34, "Alphabet Books."

- Follow the guidance on the card using *Build It From A to Z.*

Option 2: Alphabet Books and Tools

- Review Intentional Teaching Card LL34, "Alphabet Books."

- Follow the guidance on the card using *Build It From A to Z* and several labeled tools.

> Knowledge of the alphabet involves more than reciting "The ABC Song" or recognizing individual letters. Children must understand that a letter represents one or more sounds. They also need to understand that these symbols can be grouped together to form words and that words have meaning. This understanding that written spellings correspond to speech sounds is called the *alphabetic principle*. It is a key predictor of future reading success.

Mighty Minutes™

- Use Mighty Minutes 22, "Hot or Cold 3-D Shapes." Follow the guidance on the card.

Large-Group Roundup

- Recall the day's events.

- Read aloud and discuss the list of tools from large-group time.

Who builds buildings?
What tools do they use?

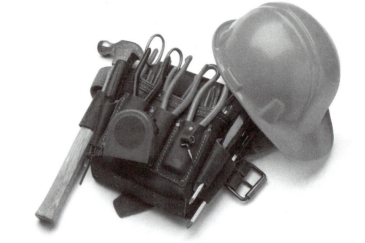

Vocabulary

English: *blueprint*

Spanish: *plano*

Large Group

Opening Routine

- Sing a welcome song and talk about who's here.

Song: "Pound the Nail"

- Use Mighty Minutes 43, "Bouncing Big Brown Balls."

- Try the carpentry version on the back of the card, e.g., pound the nail, twist the screw, or paint the wall. Try other ideas as well based on what you have done in the classroom.

Discussion and Shared Writing: Building Jobs

- Review the question of the day.

- Review the list of people who work on buildings that you created on day 1 of this investigation.

- Walk around the room with the children. Invite them to think about the people who may have worked on the different parts of the room. Say, for example, "I see some pipes that carry water to our water fountain. I wonder who might have installed those in our room. Hilary said maybe a plumber like her uncle put them in."

- Record children's ideas.

English-language learners

You may notice children using English along with words from their home language(s) in their responses. Combining words from both languages, or code switching, is to be expected and does not mean the children are confused.

Before transitioning to interest areas, explain that everyone who worked on the school building was following a plan or *blueprint*. Talk about the blueprints, clipboards, paper, and pencils in the Block area. Discuss how children may use these tools to create plans for their own buildings.

Choice Time

As you interact with children in the interest areas, make time to

- Invite children to plan on paper before they begin constructing buildings in the Block area.

- Remind them to look back at their plans as they build.

- Take photos of children's buildings and display them alongside their plans.

Read-Aloud

Read *Changes, Changes.*

- **Before you read**, ask "What do you remember about this book? Why do you think it is called *Changes, Changes*?"

- **As you read**, invite children to tell the story.

- **After you read**, flip back through the book and record children's dictation on sticky notes. Adhere them to each page as children describe what is happening on it. Reread their words to retell the story as you turn the pages of the book.

Small Group

Option 1: Show Me Five

- Review Intentional Teaching Card M16, "Show Me Five."

- Follow the guidance on the card using small building-related manipulatives, e.g., nuts and bolts, wood scraps, small tiles, or small tools.

Option 2: Guessing Jar

- Review Intentional Teaching Card M17, "Guessing Jar."

- Follow the guidance on the card using small building-related manipulatives, e.g., nuts and bolts, wood scraps, small tiles, or small tools.

Mighty Minutes™

- Use Mighty Minutes 60, "The Name Dance." Try the syllable deletion variation on the back of the card.

Large-Group Roundup

- Recall the day's events.

- Invite children who worked in the Block area to share their building plans and photos or drawings of their work.

Who builds buildings?
What tools do they use?

Vocabulary

English: *hard hat, protect, safety precaution*

Spanish: *casco, proteger, precauciones para mantenerse seguros*

See Book Discussion Card 19, *The Three Little Javelinas* (*Los tres pequeños jabalíes*), for additional words.

Large Group

Opening Routine

• Sing a welcome song and talk about who's here.

Song: "What Is My Job?"

• Use Mighty Minutes 11, "What Is My Job?"

• Follow the guidance on the card using construction-related jobs.

Discussion and Shared Writing: Builders Staying Safe

• Review the question of the day.

• Remind children about any safety precautions that the visitor talked about during the interview on day 2 of this investigation.

• Explain, "People who work on buildings have to do special things, like wearing *hard hats,* to keep their bodies safe. They take other *safety precautions*, too. I wonder what else they do to *protect* their bodies. Let's look through this book and see if we can find out."

• Look through *Build It From A to Z* with the children.

• Invite children to identify other safety equipment, e.g, protective eyewear, boots, or work gloves.

• Record children's ideas.

Before transitioning to interest areas, talk about the small construction vehicles in the Sand and Water area and how children may use them.

Choice Time

As you interact with children in the interest areas, make time to

• Record what children say and do in the interest areas.

• Ask children questions about the construction vehicles in the Sand and Water area, e.g., "How does the bucket on the front loader work? Which vehicle do you think a building crew might use to smooth out the ground before they build the house?"

> **For ideas on supporting children who have difficulty playing with others, see Intentional Teaching Card SE14, "Playing Together."**

Read-Aloud

Read *The Three Little Javelinas.*

• Review Book Discussion Card 19, *The Three Little Javelinas.* Follow the guidance for the first read-aloud.

Small Group

Option 1: Cube Trains

• Review Intentional Teaching Card M40, "Cube Trains." Follow the guidance on the card.

Option 2: Action Patterns

• Review Intentional Teaching Card M35, "Action Patterns." Follow the guidance on the card.

> For more information on helping children learn about patterns, see *The Creative Curriculum for Preschool, Volume 4: Mathematics.*

Mighty Minutes™

• Use Mighty Minutes 59, "Clap the Beat."

Large-Group Roundup

• Recall the day's events.

• Invite children who played with the construction vehicles in the Sand and Water area to share what they discovered.

Investigation 3

What are buildings made of? What makes them strong?

	Day 1	Day 2	Day 3
Interest Areas	Discovery: sample building materials, e.g., brick, wood, steel, and cinder block; magnifying glasses Computer: eBook version of *Build It From A to Z*	Toys and Games: collections of building samples, e.g., paint sample cards, carpet sample books, and tiles	Art: craft sticks; molding clay Computer: eBook version of *Build It From A to Z*
Question of the Day	What can you do with these? (Display a few different building materials.)	What do you think most of the buildings we'll see today are made of? (Display a few different building materials.)	Which house is the strongest? (Display pictures of brick, straw, and stick houses.)
Large Group	Song: "Bouncing Big Brown Balls" **Discussion and Shared Writing:** What Are Buildings Made Of? **Materials:** Mighty Minutes 43, "Bouncing Big Brown Balls"; sample building materials; *Build It From A to Z*	Game: Walk Around the Shapes **Discussion and Shared Writing:** Preparing for the Site Visit **Materials:** Mighty Minutes 52, "Walk Around the Shapes"; shape or letter cards; photos of local buildings; building materials from question of the day; Intentional Teaching Card LL45, "Observational Drawing"	**Movement:** My Body Jumps **Discussion and Shared Writing:** Strong Buildings **Materials:** Mighty Minutes 72, "My Body Jumps"
Read-Aloud	*The Three Little Javelinas* Book Discussion Card 19 (second read-aloud)	*Building a House*	*Build It From A to Z*
Small Group	**Option 1: Geoboards** Intentional Teaching Card M21, "Geoboards"; geoboards; shape cards; geobands **Option 2: Straw Shapes** Intentional Teaching Card M42, "Straw Shapes"; geometric shapes; drinking straws of different lengths; pipe cleaners; paper; pencils or crayons	**Option 1: Can We Build It Together?** Intentional Teaching Card SE25, "What Can We Build Together?"; building blocks **Option 2: Can We Build It Together? Construction Sounds** Intentional Teaching Card SE25, "What Can We Build Together?"; building blocks; variety of building materials	**Option 1: I'm Thinking of a Shape** Intentional Teaching Card M20, "I'm Thinking of a Shape"; geometric solids; empty containers that have basic shapes **Option 2: Buried Shapes** Intentional Teaching Card M30, "Buried Shapes"; heavy paper; attribute blocks; three containers to use as treasure chests; tub of sand; small brushes; book about a treasure (optional)
Mighty Minutes™	Mighty Minutes 60, "The Name Dance"	Mighty Minutes 19, "I Spy With My Little Eye"	Mighty Minutes 50, "1, 2, 3, What Do I See?"; small basket of building-related items; scarf or piece of fabric

Day 4	Day 5	Make Time For...
Block: unit blocks; other building materials **Discovery:** stick houses that the children made yesterday in the Art area	**Block:** big boxes; masking tape	## Outdoor Experiences **Continuing to Observe a Building's Shadow** • Invite children to observe the school's shadow at a different time of day from last week's observation.
Is this building sturdy? (Display a tall block tower.)	Can we make a building out of this? (Display a cardboard box.)	• Have them trace around the shadow with sidewalk chalk. • Repeat this process at different times of the day for a few days. • Talk about how the shadow's size and position are different from those observed during previous observations.
Movement: I Can Make a Circle **Discussion and Shared Writing:** Sturdy Buildings **Materials:** Mighty Minutes 20, "I Can Make a Circle"; small blocks; sticky notes; crayons	**Poem:** "A Building My Size" **Discussion and Shared Writing:** Cardboard Buildings **Materials:** Mighty Minutes 49, "A Tree My Size"; large cardboard boxes	**Physical Fun** • Use Intentional Teaching Card P22, "Follow the Leader." Follow the guidance on the card. ## Family Partnerships • Ask families to bring in large, empty boxes for children to use to construct buildings at the end of the investigation.
The Three Little Javelinas Book Discussion Card 19 (third read-aloud)	*Building a House*	## Wow! Experiences • Day 2: A walk around the neighborhood to investigate the materials used to construct neighborhood buildings and identify problems
Option 1: Which Container Holds More? Intentional Teaching Card M32, "Which Container Holds More?"; sand table or tubs of sand; containers of various sizes; funnel; paper; marker; paper cup, measuring cup, or can **Option 2: Cover Up** Intentional Teaching Card M34, "Cover Up"; masking tape; samples and pictures of various floor coverings; blocks	**Option 1: Tongue Twisters** Intentional Teaching Card LL16, "Tongue Twisters" **Option 2: Same Sound Sort** Intentional Teaching Card LL12, "Same Sound Sort"; objects that start with the /b/ sound; cardboard box	
Mighty Minutes 50, "1, 2, 3, What Do I See?"; small basket of building-related items; scarf or piece of fabric	Mighty Minutes 13, "Simon Says"	

What are buildings made of?
What makes them strong?

Vocabulary

See Book Discussion Card 19, *The Three Little Javelinas (Los tres pequeños jabalíes)*.

Large Group

Opening Routine

• Sing a welcome song and talk about who's here.

Song: "Bouncing Big Brown Balls"

• Use Mighty Minutes 43, "Bouncing Big Brown Balls."

• Try the variation on the back of the card using building-related actions.

Discussion and Shared Writing: What Are Buildings Made Of?

• Review the question of the day.

• Show the examples of building materials that will be in the Discovery area.

• Talk about each material as you pass it around for children to explore.

• Invite them to test the strength of the materials in different ways, such as bending, stretching, or standing on them.

• Explain, "Buildings are made from lots of different materials."

• Reread *Build It From A to Z.* Call children's attention to the different building materials. Record the names of the materials as you read about them.

Before transitioning to interest areas, talk about the building materials in the Discovery area and how children may use them. Tell the children that the book will be available on the computer in the Computer area.

Choice Time

As you interact with children in the interest areas, make time to

• Talk to children as they explore the building materials.

• Ask questions to help them consider the materials' use and strength.

• Invite them to test the strength of individual materials.

English-language learners

When you think children are ready, begin challenging them to use English instead of nonverbal techniques to communicate. For example, if a child requests nonverbally that you retrieve a tool from a shelf, point to the shelf and ask, "Do you want something from this shelf? Do you want a hammer?" When they are ready, children begin using words or phrases to communicate their needs.

Read-Aloud	Read *The Three Little Javelinas*. • Use Book Discussion Card 19, *The Three Little Javelinas*. Follow the guidance for the second read-aloud.

Small Group

Option 1: Geoboards

• Review Intentional Teaching Card M21, "Geoboards." Follow the guidance on the card.

Option 2: Straw Shapes

• Review Intentional Teaching Card M42, "Straw Shapes." Follow the guidance on the card.

Mighty Minutes™

• Use Mighty Minutes 60, "The Name Dance." Try the syllable deletion variation on the back of the card.

Large-Group Roundup

• Recall the day's events.

• Invite children who explored building materials in the Discovery area to describe their findings.

• Ask, "What would you like to make buildings out of? What are some things we can use to make buildings? How do you think we could make a strong building?"

• Record children's responses.

> Try to obtain as many of the materials the children identified as you can. Having a variety of building materials will encourage the children to spend the week constructing and testing their buildings. Encourage children to draw plans for their buildings before they begin construction. Document their discoveries with photographs and their observational drawings.

What are buildings made of?
What makes them strong?

Vocabulary

English: *foundation, building inspector*

Spanish: *cimientos, inspector(a) de edificios*

Large Group

Opening Routine

- Sing a welcome song and talk about who's here.

Game: Walk Around the Shapes

- Use Mighty Minutes 52, "Walk Around the Shapes." Follow the guidance on the card.

Discussion and Shared Writing: Preparing for the Site Visit

- Review the question of the day.

- Pass around some photos of neighborhood buildings that you took during the first investigation.

- Invite children to describe what each building looks like and what it is made of.

- Pass around the building materials from the question of the day.

- Ask, "Can you tell what materials were used to make the buildings in the pictures? Were any of these materials used?"

- Record children's responses.

- Explain, "After today's walk around the neighborhood, we can check our predictions to see whether we were right."

- Introduce the idea of building inspection. Explain, "Sometimes buildings have problems. It is the *building inspector's* job to make sure that a building stays strong and safe. On today's walk, let's see whether we can identify things a building inspector would check."

- See Intentional Teaching Card LL45, "Observational Drawing."

> **During the walk, help children identify and name building materials, such as asphalt or roofing shingles. Point out problems if the children don't notice them easily, such as peeling paint and broken windows.**

Before transitioning to interest areas, talk about the collection of building samples, such as carpet or tiles, in the Toys and Games area, and how children may use them. Show children the materials that you gathered on the basis of the suggestions they made during yesterday's large-group roundup. Add the materials to an appropriate interest area. Explain, "You may use these materials to construct buildings and test how strong they are."

Choice Time

As you interact with children in the interest areas, make time to

- Talk with children about the building samples in the Toys and Games area.

- Use rich vocabulary to describe the materials' textures, styles, and designs, e.g., say, "This carpet feels silky," or "This tile has a plaid pattern."

Read-Aloud

Read *Building a House*.

- **Before you read**, ask, "What do you remember about this book?"

- **As you read**, add details to the text on each page and use vocabulary terms, such as *foundation,* to expand on the information in the book. Say, for example, "The cement mixer pours the

cement. This will be the *foundation* for the building. The *foundation* provides a strong support to build on."

- **After you read**, look through the pages of the book with the children. Talk about and list the variety of materials used to construct the building.

Small Group

Option 1: What Can We Build Together?

- Review Intentional Teaching Card SE25, "What Can We Build Together?" Follow the guidance on the card.

Option 2: What Can We Build Together? Construction Sounds

- Review Intentional Teaching Card SE25, "What Can We Build Together?"

- Provide a variety of building materials. Follow the guidance on the card.

Mighty Minutes™

- Use Mighty Minutes 19, "I Spy With My Little Eye." Try the shape variation on the back of the card.

Large-Group Roundup

- Recall the day's events.
- Talk about what children discovered during today's site visit.

- Record their ideas.
- Invite children to show their observational drawings.

What are buildings made of?
What makes them strong?

Vocabulary

English: *sturdy, foundation*

Spanish: *resistente, cimientos*

Large Group

Opening Routine

- Sing a welcome song and talk about who's here.

Movement: My Body Jumps

- Use Mighty Minutes 72, "My Body Jumps." Follow the guidance on the card.

Discussion and Shared Writing: Strong Buildings

- Say, "On our walk yesterday, we saw lots of strong, *sturdy* buildings in our neighborhood." Remind children about the problem that the three little pigs faced in the story *The Three Little Pigs*.

- Review the question of the day.

- Wonder aloud, "How could the first two little pigs have made their houses stronger?"

- Record children's ideas.

Before transitioning to interest areas, talk about the craft sticks and clay in the Art area and how children may use them together to make a sturdy house out of sticks.

Choice Time

As you interact with children in the interest areas, make time to

- Talk to the children as they build stick houses in the Art area.

- Ask questions that encourage children to describe their process as they work: "Why did you put the stick there? What will you add next? How will you make your building *sturdy*?"

English-language learners
Focused, one-on-one conversations about, for example, a child's choice of building material make it easier for English-language learners to communicate because the contexts are narrow. Such conversations will also help you understand what the child is trying to say and enable you to teach particular vocabulary and concepts.

Read-Aloud	Read *Build It From A to Z.* • **Before you read**, explain, "I will need your help to read this book today." • **As you read**, invite children to name the letters and any of the words they remember from the book.	• **After you read**, flip through the book and say children's names as you show the pages that correspond to the first letter in their names. For example, say "*Abraham* and *Alia* both began with an A. Tell the children that the book will be available to them on the computer in the Computer area.
Small Group	**Option 1: I'm Thinking of a Shape** • Review Intentional Teaching Card M20, "I'm Thinking of a Shape." Follow the guidance on the card.	**Option 2: Buried Shapes** • Review Intentional Teaching Card M30, "Buried Shapes." Follow the guidance on the card.
Mighty Minutes™	• Use Mighty Minutes 50, "1, 2, 3, What Do I See?" Follow the guidance on the card.	
Large-Group Roundup	• Recall the day's events. • Invite children who built stick houses in the Art area to share their work. • Ask, "How can we test these buildings to find out whether they're *sturdy*?"	• Record children's ideas. • Say, "Tomorrow you may try some of these ideas to test your buildings in the Discovery area."

What are buildings made of?
What makes them strong?

Vocabulary

English: *sturdy*

Spanish: *resistente*

See Book Discussion Card 19, *The Three Little Javelinas*
(Los tres pequeños jabalíes), for additional words.

Large Group

Opening Routine

- Sing a welcome song and talk about who's here.

Movement: I Can Make a Circle

- Use Mighty Minutes 20, "I Can Make a Circle." Follow the guidance on the card.

Discussion and Shared Writing: Sturdy Buildings

- Review the question of the day.

- Begin stacking small blocks. Ask, "How many blocks tall do you think we can build this tower before it falls?"

- Have children record their predictions on sticky notes or make predictions as a group.

- Count the blocks as you and the children stack them.

- After they fall, say, "I noticed that you are building a lot of interesting tall buildings. Tall buildings are a challenge to build because it is hard to make them *sturdy* so they won't fall down."

- Ask, "What tall buildings have you seen? What makes them strong?"

- Record children's responses.

Before transitioning to interest areas, say, "Let's experiment in the Block area today with what makes our block buildings *sturdy*." Review the ideas that children generated yesterday for testing the sturdiness of their stick houses. Invite children who are interested to test their houses in the Discovery area.

Choice Time

As you interact with children in the interest areas, make time to

- Talk to children as they build in the Block area.

- Invite them to try different building techniques. Ask, "Will that make the building *sturdy*?"

- Invite children to test the strength of their stick houses in the Discovery area. Encourage children to make predictions, test the houses, and compare their findings with their predictions.

Read-Aloud

Read *The Three Little Javelinas.*

- Use Book Discussion Card 19, *The Three Little Javelinas.* Follow the guidance for the third read-aloud.

Small Group

Option 1: Which Container Holds More?

- Review Intentional Teaching Card M32, "Which Container Holds More?" Follow the guidance on the card.

Option 2: Cover Up

- Review Intentional Teaching Card M34, "Cover Up." Follow the guidance on the card.

Mighty Minutes™

- Use Mighty Minutes 50, "1, 2, 3, What Do I See?" Follow the guidance on the card.

Large-Group Roundup

- Recall the day's events.
- Invite the children who worked in the Block area during choice time to share what they discovered about building sturdy structures.
- Ask, "What are the rules for creating *sturdy* buildings?"

- Record children's responses, which might include, "Put heavy blocks on the bottom and little blocks on the top," and, "Make the building wider at the bottom than at the top."

> **See Intentional Teaching Card LL43, "Introducing New Vocabulary" for ways to support expanding a child's vocabulary.**

What are buildings made of?
What makes them strong?

Vocabulary

English: *characteristics*

Spanish: *características*

Large Group

Opening Routine

- Sing a welcome song and talk about who's here.

Poem: "A Building My Size"

- Use Mighty Minutes 49, "A Tree My Size."

- Follow the guidance on the card and substitute the word *building* for *tree*.

- Invite the children to jump or touch a different part of their body as each word is read to create a movement pattern for the four-word sentences, e.g., for *this*, touch your feet; for *building*, touch your knees; for *is*, touch your shoulders; and for *tall*, touch your head.

> **Encouraging children to make a movement for each word helps them separate a sentence into individual words.**

Discussion and Shared Writing: Cardboard Buildings

- Review the children's findings about materials used to make buildings strong and sturdy.

- Show the children a few large cardboard boxes.

- Review the question of the day.

- Say, "When I saw these cardboard boxes, I thought about how much fun it would be to make them into buildings for our classroom! But I need your help to do it."

- Ask, "How could we use these boxes to make a sturdy pretend building?"

- Record children's ideas.

Before transitioning to interest areas, talk about the big boxes and masking tape in the Block area and how children may use them to make buildings.

Choice Time

As you interact with children in the interest areas, make time to

- Invite children to discuss or draw a building plan before they start to build with the boxes and masking tape.

- When asked, help children cut holes in the boxes for doors and windows.

- Discuss the *characteristics* of the children's buildings.

Read-Aloud

Read *Building a House*.

- **Before you read**, remind the children that this book describes the steps for building a house.

- **As you read**, invite the children to compare the steps in the book to the process they used to construct buildings out of large cardboard boxes.

- **After you read**, ask, "What did the workers do to make sure the house they built was sturdy?"

Small Group

Option 1: Tongue Twisters

- Review Intentional Teaching Card LL16, "Tongue Twisters."

- Follow the guidance on the card, using the following tongue twister:

  Busy workers bravely balance banging boards on the building's beams.

Option 2: Same Sound Sort

- Review Intentional Teaching Card LL12, "Same Sound Sort."

- Follow the guidance on the card using items that start with the /b/ sound.

English-language learners
English-language learners may have difficulty saying tongue twisters if they include sounds that are not found in their home languages. As children learn to produce sounds specific to English, accept and acknowledge their attempts without correcting errors. However, continue to model correct pronunciation. If possible, make up a tongue twister with a beginning sound that the children can pronounce easily.

Mighty Minutes™

- Use Mighty Minutes 13, "Simon Says."

- Follow the guidance on the card, using the characteristics of objects, e.g., say, "Touch something green (or tall, or sturdy)."

Large-Group Roundup

- Recall the day's events.

- Review the /b/ sound and invite children to name words that start with it.

Investigation 4

What is special about our building?

	Day 1	Day 2	
Interest Areas	Art: small cardboard boxes; empty milk cartons; construction paper; paint; tape; glue; scissors; neighborhood map from Investigation 1	Art: small cardboard boxes; empty milk cartons; construction paper; paint; tape; glue; scissors; neighborhood map from Investigation 1	
Question of the Day	How many classrooms do you think are in our school?	Where in our school is *this*? (Display a picture of the feature you will visit today.)	
Large Group	**Game:** Hot or Cold 3-D Shapes **Discussion and Shared Writing:** Our School Building **Materials:** Mighty Minutes 22, "Hot or Cold 3-D Shapes"; several three-dimensional shapes; tally sheets; photos of neighborhood buildings	**Game:** Riddle, Riddle, What Is That? **Discussion and Shared Writing:** Preparing for the Site Visit **Materials:** Mighty Minutes 61, "Riddle, Riddle, What Is That?"; Intentional Teaching Card LL45, "Observational Drawing"	
Read-Aloud	*The True Story of the 3 Little Pigs* Book Discussion Card 22 (first read-aloud)	*Changes, Changes* building blocks	
Small Group	**Option 1: Stick Letters** Intentional Teaching Card LL28, "Stick Letters"; collection of sticks; alphabet cards **Option 2: Walk a Letter** Intentional Teaching Card LL17, "Walk a Letter"; masking tape; alphabet cards or alphabet chart	**Option 1: The Long & Short of It** Intentional Teaching Card M25, "The Long & Short of It"; pieces of yarn or ribbon of equal width but different lengths; a container **Option 2: Lining It Up** Intentional Teaching Card M31, "Lining It Up"; collection of objects to be arranged by size	
Mighty Minutes™	Mighty Minutes 53, "Three Rowdy Children"	Mighty Minutes 04, "Riddle Dee Dee"	

Day 3

Art: small cardboard boxes; empty milk cartons; construction paper; paint; tape; glue; scissors; neighborhood map from Investigation 1

What would you like to ask our visitor today?

Game: Clap the Beat

Discussion and Shared Writing: Expert Interview

Materials: Mighty Minutes 59, "Clap the Beat"; several building-related items or photographs

The True Story of the 3 Little Pigs
Book Discussion Card 22
(second read-aloud)

Option 1: Knowing Our Friends

Intentional Teaching Card LL30, "Knowing Our Friends"; children's name cards; felt board or tagboard; large paper clip or Velcro®

Option 2: What's for Snack?

Intentional Teaching Card LL25, "What's for Snack?"; food labels; large paper or tagboard; marker; recipe cards or charts

Mighty Minutes 07, "Hippity, Hoppity, How Many?"

Make Time For...

Outdoor Experiences

Physical Fun

- Use Intentional Teaching Card P17, "Balance on a Beam." Follow the guidance on the card.

Family Partnerships

- Invite family members to accompany the class on the site visit to the neighborhood building while exploring the investigation question for day 2.

Wow! Experiences

- Day 1: A walk around the inside of the school building to tally the different types of rooms
- Day 2: An in-depth investigation of an interesting feature of the school, e.g., stairs, elevator, fire escape, atrium, or gymnasium
- Day 3: A visit from someone, such as a maintenance worker, handyman, or electrician, who helps maintain the building

What is special about our building?

Vocabulary

English: *model*

Spanish: *modelo*

See Book Discussion Card 22, *The True Story of the 3 Little Pigs* (*La verdadera historia de los tres cerditos*), for additional words.

Large Group

Opening Routine

- Sing a welcome song and talk about who's here.

Game: Hot or Cold 3-D Shapes

- Use Mighty Minutes 22, "Hot or Cold 3-D Shapes." Follow the guidance on the card.

Discussion and Shared Writing: Our School Building

- Explain, "We know so much about how buildings are made. Now let's closely examine our school building."

- Review the question of the day.

- Ask, "What are some of the other rooms that we have in our school building?"

- Record children's responses.

- Explain, "Today we will take a walk around our school and count the different kinds of rooms that we find."

> Create tally sheets in advance for the children to use to record the number of different rooms in the school.

Before transitioning to interest areas, look at some of the photos of the neighborhood buildings that you've taken throughout the study. Remind children of interesting observations that they've made about the buildings while exploring previous investigation questions. Explain, "We are going to work in the Art area during the next few days to make a *model* of our neighborhood." Talk about the cardboard construction materials in the Art area and how children may use them to make models of neighborhood buildings.

Choice Time

As you interact with children in the interest areas, make time to

- Invite children to look at the neighborhood map that you created for Investigation 1.

- Invite children to select a photograph of a neighborhood building that they'd like to re-create.

- Ask, "What can you tell me about the building in the picture you selected? What do you like about this building? What materials will you use to make a *model* of it?"

Read-Aloud

Read *The True Story of the 3 Little Pigs.*

- Use Book Discussion Card 22, *The True Story of the 3 Little Pigs.* Follow the guidance for the first read-aloud.

Small Group

Option 1: Stick Letters

- Review Intentional Teaching Card LL28, "Stick Letters." Follow the guidance on the card.

Option 2: Walk a Letter

- Review Intentional Teaching Card LL17, "Walk a Letter." Follow the guidance on the card.

Mighty Minutes™

- Use Mighty Minutes 53, "Three Rowdy Children." Follow the guidance on the card.

Large-Group Roundup

- Recall the day's events.
- Talk about today's walk around the inside of the school.
- Review the question of the day. Compare children's guesses to the actual number of classrooms tallied on the walk.
- Use a Venn diagram to show which rooms are found in a home and school. Identify rooms that are unique and the same in both buildings. For example, a bathroom is found in both places, while a bedroom is found in a home, and a janitor's office is found in a school.

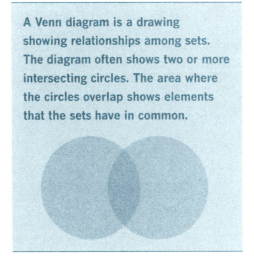

A Venn diagram is a drawing showing relationships among sets. The diagram often shows two or more intersecting circles. The area where the circles overlap shows elements that the sets have in common.

What is special about our building?

Vocabulary

English: *feature*

Spanish: *característica*

Large Group

Opening Routine

- Sing a welcome song and talk about who's here.

Game: Riddle, Riddle, What Is That?

- Use Mighty Minutes 61, "Riddle, Riddle, What Is That?"
- Follow the guidance on the card and describe various construction tools and equipment.

Discussion and Shared Writing: Preparing for the Site Visit

- Review what you discovered together during yesterday's walk around the school.
- Review the question of the day.
- Ask questions to encourage discussion about the special feature that you'll be studying today, e.g., "What is this used for? In what other buildings have you seen something like this?"

- Record children's responses.
- Review Intentional Teaching Card LL45, "Observational Drawing." Follow the guidance on the card, encouraging children to create observational drawings on the site visit.

> While investigating the special feature on the site visit, encourage children to hypothesize how different parts of it were constructed. Ask, "How do you think the builders made the doorways? the corners? the curved lines?"

Before transitioning to interest areas, remind children about the materials in the Art area. Explain that children may use them to construct models of neighborhood buildings.

Choice Time

As you interact with children in the interest areas, make time to

- Invite children to refer back to their selected building photographs as they work.

- Ask questions that encourage them to notice specific details of the buildings in their photographs.

Read-Aloud

Read *Changes, Changes.*

- **Before you read**, have the children find partners. Give each pair a small set of building blocks.

- **As you read**, pause occasionally and invite children to build the different structures in the story.

- **After you read**, talk to the children about the fire in the story. Ask, "Do you know what would keep us and our building safe if there were a fire?" Point out any fire safety features in your classroom, e.g., evacuation plan, fire extinguisher, fire alarm, smoke detector, and sprinkler.

Small Group

Option 1: The Long & Short of It

- Review Intentional Teaching Card M25, "The Long & Short of It." Follow the guidance on the card.

Option 2: Lining It Up

- Review Intentional Teaching Card M31, "Lining It Up." Follow the guidance on the card.

Mighty Minutes™

- Use Mighty Minutes 04, "Riddle Dee Dee." Try the syllable jumping variation on the back.

Large-Group Roundup

- Recall the day's events.

- Talk about today's site visit. Ask, "What did you notice about the special feature of our school?"

- Record children's observations.

- Invite children to share their observational drawings.

- Tell the children that tomorrow someone will be visiting the class to talk about how she helps take care of the school.

- Ask, "What would you like to ask our visitor tomorrow?"

- Record children's questions.

What is special about our building?

Vocabulary

See Book Discussion Card 22, *The True Story of the 3 Little Pigs* (*La verdadera historia de los tres cerditos*).

Large Group

Opening Routine

- Sing a welcome song and talk about who's here.

Game: Clap the Beat

- Use Mighty Minutes 59, "Clap the Beat." Follow the guidance on the card.

Discussion and Shared Writing: Expert Interview

- Introduce the visitor.

- Invite the visitor to talk about what he or she does to maintain the school or repair the buildings, e.g., a broken window, leaky faucet, loose nail, or chipped paint.

- Invite the children to ask the questions they came up with in yesterday's large-group roundup and the question of the day.

- Invite the visitor to repair something in the classroom. If nothing needs repair, ask her to show you what would be done in the event of a specific problem. For example, ask, "Can you show us what you would do if the faucet leaked? What tools would you use? What would you look at first?"

Before transitioning to interest areas, remind children about the materials in the Art area. Explain that they may use them to construct models of neighborhood buildings. Encourage children to look back at the observational drawings they made yesterday. Challenge the children to find materials in the classroom that can be used to construct models of what they recorded in their observational drawings.

Choice Time

As you interact with children in the interest areas, make time to

- Invite children to help you display the model neighborhood buildings that they created in the Art area.

- Remind children about the neighborhood map. Explain, "We can use the map to help us decide where to put the buildings in our display."

- Invite the children to add other features to the display, such as roads, parks, and traffic signals.

English-language learners
When children are beginning to speak in sentences, do not correct their grammar. For all children, model the correct use of English. For example, if a child says, "They is going to store," you might reword the statement and then ask a question to encourage more discussion. Say, "They are going to the store. What will they buy there?"

Read-Aloud

Read *The True Story of the 3 Little Pigs*.

- Use Book Discussion Card 22, *The True Story of the 3 Little Pigs*. Follow the guidance for the second read-aloud.

Small Group

Option 1: Knowing Our Friends

- Review Intentional Teaching Card LL30, "Knowing Our Friends." Follow the guidance on the card.

Option 2: What's for Snack?

- Review Intentional Teaching Card LL25, "What's for Snack?" Follow the guidance on the card.

Mighty Minutes™

- Use Mighty Minutes 7, "Hippity, Hoppity, How Many?" Follow the guidance on the card.

Large-Group Roundup

- Recall the day's events.
- Invite children who helped with the model neighborhood display in the Art area to share any special features that they added to it.

- Write a group thank-you note to the visitor who repairs buildings. Invite children to sign their names and add drawings to the note.

Investigation 5

What happens inside buildings?

	Day 1	Day 2	Day 3
Interest Areas	**Sand and Water:** forms and containers to mold wet sand **Computer:** eBook version of *Buildings, Buildings, Buildings*	**Sand and Water:** additional forms and containers to mold wet sand	**Discovery:** nuts and bolts
Question of the Day	What do you think people do in this building? (Display a picture of an interesting building.)	What would you like to ask our visitor today?	Do you see any letters that you recognize on this building sign? (Display a picture of a building sign.)
Large Group	**Song:** "La, La, La" **Discussion and Shared Writing:** What's Going On Inside? **Materials:** Mighty Minutes 100, "La, La, La"; pictures of building exteriors; picture of a neighborhood building	**Game:** People Patterns **Discussion and Shared Writing:** Interviewing a Neighbor **Materials:** Mighty Minutes 65, "People Patterns"; picture of a neighborhood building	**Song:** "The People in Your Neighborhood" **Discussion and Shared Writing:** Preparing for the Site Visit **Materials:** Mighty Minutes 01, "The People in Your Neighborhood"; picture of the building that you will see
Read-Aloud	*Buildings, Buildings, Buildings*	*The True Story of the 3 Little Pigs* Book Discussion Card 22 (third read-aloud)	*The Pot That Juan Built*
Small Group	**Option 1: Secret Numbers** Intentional Teaching Card M37, "Secret Numbers"; two sets of quantity cards, numeral–quantity cards, or numeral cards **Option 2: Making Numerals** Intentional Teaching Card M41, "Making Numerals"; *Keep Counting*; modeling dough or clay; numeral–quantity cards	**Option 1: Patterns** Intentional Teaching Card M14, "Patterns"; group of objects to be arranged in a pattern; examples of patterns; construction paper **Option 2: Patterns Under Cover** Intentional Teaching Card M38, "Patterns Under Cover"; counters in a variety of colors; paper cup; cardboard divider	**Option 1: Playing With Environmental Print** Intentional Teaching Card LL23, "Playing With Environmental Print"; environmental print; pictures of road or store signs **Option 2: Coupon Match** Intentional Teaching Card LL22, "Coupon Match"; empty product containers or food labels; laminated coupons that correspond to products; grocery bag; envelope
Mighty Minutes™	Mighty Minutes 59, "Clap the Beat"; several building-related items or photographs	Mighty Minutes 55, "Mr. Forgetful"	Mighty Minutes 36, "Body Patterns"

Day 4	Day 5	Make Time For…
Dramatic Play: props that reflect the inside of the neighborhood building that you visited	**Dramatic Play:** props that reflect the inside of the neighborhood building that you visited	## Outdoor Experiences
Computer: eBook version of *Buildings, Buildings, Buildings*	**Computer:** eBook version of *The Three Little Pigs*	**Physical Fun** • Use Intentional Teaching Card P14, "Moving Through the Forest." Follow the guidance on the card.
What do you think people do in this building? (Display a picture of a familiar building with a pictorial sign that shows what happens in the building, e.g., a car wash sign with a soapy car.)	Which book would you like to read today? (Display the three versions of the story about the three little pigs.)	## Family Partnerships • Invite families to attend the end-of-study celebration. Send home a letter explaining the event and listing the ideas children have for fixing things around the school building.
Movement: Counting Calisthenics	**Movement:** Bounce, Bounce, Bounce	## Wow! Experiences
Discussion and Shared Writing: Building Signs	**Discussion and Shared Writing:** Three Versions of the Same Story	• Day 2: Visit from a neighbor who knows about a neighborhood building
Materials: Mighty Minutes 28, "Counting Calisthenics"; pictures of building signs that give clues about their purpose	**Materials:** Mighty Minutes 30, "Bounce, Bounce, Bounce"; *The Three Little Pigs*, *The Three Little Javelinas*, and *The True Story of the 3 Little Pigs*	• Day 3: Site visit to look inside a different neighborhood building and learn about what people do there
Buildings, Buildings, Buildings	*The Three Little Pigs*, *The Three Little Javelinas*, or *The True Story of the 3 Little Pigs*	
Option 1: Bookmaking	**Option 1: Bookmaking**	
Intentional Teaching Card LL04, "Bookmaking"; pictures of interesting buildings; card stock; paper; writing tools; bookbinding supplies	Intentional Teaching Card LL04, "Bookmaking"; pictures of interesting buildings; card stock; paper; writing tools; bookbinding supplies	
Option 2: Desktop Publishing	**Option 2: Desktop Publishing**	
Intentional Teaching Card LL02, "Desktop Publishing"; building pictures; digital camera; computer; each child's word bank; printer and paper; bookbinding supplies	Intentional Teaching Card LL02, "Desktop Publishing"; interesting building pictures; digital camera; computer; each child's word bank; printer and paper; bookbinding supplies	
Mighty Minutes 36, "Body Patterns"	Mighty Minutes 38, "Spatial Patterns"	

What happens inside buildings?

Vocabulary

English: *substitute*

Spanish: *sustituir*

Large Group

Opening Routine

- Sing a welcome song and talk about who's here.

Song: "La, La, La"

- Use Mighty Minutes 100, "La, La, La." Follow the guidance on the card.

- Explain, "We are going to *substitute* 'la, la' for the words in a song."

Discussion and Shared Writing: What's Going On Inside?

- Review the question of the day.

- Show children several pictures of various building exteriors.

- Choose one of them and ask, "What do you think happens inside this building?"

- Record children's responses.

- Show a picture of a neighborhood building.

- Ask, "What do you think happens inside this building?"

- Record their responses.

- Explain, "Tomorrow we are having a visitor. He is a neighbor who knows about this building and wants to talk to us about it. We'll then be able to find out whether your predictions about what goes on in the building were right."

Before transitioning to interest areas, talk about the forms and containers in the Sand and Water area and how children may use them.

Choice Time

As you interact with children in the interest areas, make time to

- Observe children as they work with the forms and containers in the Sand and Water area.

- Ask questions to encourage children to think about the difference between building with wet sand and building with dry sand.

For information on supporting children as they make construction choices, see Intentional Teaching Card SE15, "Making Choices."

Read-Aloud	Read *Buildings, Buildings, Buildings*.
	• **Before you read**, ask, "What was this book about?"
	• **As you read**, pause to talk about what is happening inside the building.
	• **After you read**, review any of the pictures in the book that are similar to buildings in the neighborhood. Tell the children that the book will be available on the computer in the Computer area.

Small Group	• **Option 1: Secret Numbers**	**Option 2: Making Numerals**
	• Review Intentional Teaching Card M37, "Secret Numbers." Follow the guidance on the card.	• Review Intentional Teaching Card M41, "Making Numerals." Follow the guidance on the card using *Keep Counting*.

Mighty Minutes™	• Use Mighty Minutes 59, "Clap the Beat." Follow the guidance on the card.

Large-Group Roundup	• Recall the day's events.	• Ask, "What would you like to ask our neighbor about the building in this picture?"
	• Show the picture of the neighborhood building that you talked about during large-group time.	• Record children's questions.
	• Remind children that a neighbor will be visiting the class tomorrow.	

What happens inside buildings?

Vocabulary

See Book Discussion Card 22, *The True Story of the 3 Little Pigs (La verdadera historia de los tres cerditos)*.

Large Group

Opening Routine

- Sing a welcome song and talk about who's here.

Game: People Patterns

- Use Mighty Minutes 65, "People Patterns." Follow the guidance on the card.

Discussion and Shared Writing: Interviewing a Neighbor

- Show the picture of the neighborhood building from yesterday's large-group time.

- Recall the children's predictions about what happens inside that building.

- Introduce the neighbor who knows about the building.

- Invite your guest to describe what happens in the building.

- Invite children to ask the questions they generated during yesterday's large-group roundup and the question of the day.

- Record the guest's responses.

Before transitioning to interest areas, talk about the new forms and containers in the Sand and Water area and how children may use them.

> When adding new forms and containers to the Sand and Water area, be sure to remove other props. Having too many props in the area can interfere with children's play.

Choice Time

As you interact with children in the interest areas, make time to

- Ask questions to encourage children's thinking and problem solving, e.g., "Why is that sand tower still standing while the other one fell? Why did you decide to add more water to your bucket first? Is there a way to make that sand building taller?"

Read-Aloud

Read *The True Story of the 3 Little Pigs.*

- Use Book Discussion Card 22, *The True Story of the 3 Little Pigs.* Follow the guidance for the third read-aloud.

Small Group

Option 1: Patterns

- Review Intentional Teaching Card M14, "Patterns." Follow the guidance on the card.

Option 2: Patterns Under Cover

- Review Intentional Teaching Card M38, "Patterns Under Cover." Follow the guidance on the card.

Mighty Minutes™

- Use Mighty Minutes 55, "Mr. Forgetful." Follow the guidance on the card.

Large-Group Roundup

- Recall the day's events.

- Write a group thank-you note to the neighbor who visited the classroom. Invite children to sign their names and add drawings to the note.

- Explain, "Next week we will be having a celebration to share what we've learned about buildings with our families and friends. During our celebration, we will be doing work around our school to help take care of our building."

- Ask, "What do you think you and our guests could do to help take care our building during our celebration?" Responses might include "fix a broken fence post, wash the walls that face the playground, pick up litter around the school, and sweep the sidewalks."

- Record children's responses on a chart, and save it for the celebration.

Send home a letter to families including the list of chores children generated. Ask volunteers to help with specific projects or donate needed materials. Make sure to coordinate your effort with school administrators and custodial staff.

Investigation 5

What happens inside buildings?

Vocabulary

English: *potter*

Spanish: *alfarero*

Large Group

Opening Routine

- Sing a welcome song and talk about who's here.

Song: "The People in Your Neighborhood"

- Use Mighty Minutes 01, "The People in Your Neighborhood." Follow the guidance on the card.

Discussion and Shared Writing: Preparing for the Site Visit

- Gather around the models of the neighborhood buildings that the children created together during the last investigation.

- Invite the children to talk about the buildings that they know about.

- Show a picture of the neighborhood building that you will be visiting during today's site visit. Point out the model from the display of neighborhood buildings.

- Ask, "What can you tell me about this building?"

- Record children's ideas.

- Explain, "Today we will visit this building to learn more about what happens inside it."

Before transitioning to interest areas, talk about the nuts and bolts in the Discovery area and how children may use them.

Choice Time

As you interact with children in the interest areas, make time to

- Observe children's ability to manipulate the nuts and bolts in the Discovery area.

- Record your observations.

Read-Aloud

Read *The Pot That Juan Built*.

- **Before you read**, tell children the name of the book. Ask, "How do you think Juan will build his pot? What materials will he use?"

- **As you read**, invite children to chime in on the repetitive phrase, "The beautiful pot that Juan built."

- **After you read**, recall children's predictions and discuss whether they were correct. Explain that this book is about a real person, Juan Quezada, and that he is a *potter* who lives in Mexico. Briefly share any additional information from the explanatory pages you think might be interesting to the children.

Small Group

Option 1: Playing With Environmental Print

- Talk about the question of the day.

- Review Intentional Teaching Card LL23, "Playing With Environmental Print." Follow the guidance on the card.

Option 2: Coupon Match

- Talk about the question of the day.

- Review Intentional Teaching Card LL22, "Coupon Match." Follow the guidance on the card.

- Explain the word *coupon*.

English-language learners
Include children's home languages as you collect print samples. This helps English-language learners feel proud of their culture and families. It also helps English-speaking children see that English-language learners can participate in activities.

Mighty Minutes™

- Use Mighty Minutes 36, "Body Patterns." Follow the guidance on the card.

Large-Group Roundup

- Recall the day's events.

- Encourage children to talk about today's site visit to the neighborhood building.

- Record their discoveries.

What happens inside buildings?

Vocabulary

English: *clue, calisthenics*

Spanish: *indicio, calistenia*

Large Group

Opening Routine

- Sing a welcome song and talk about who's here.

Movement: Counting Calisthenics

- Use Mighty Minutes 28, "Counting Calisthenics." Follow the guidance on the card.

- Introduce the activity by explaining that *calisthenics* is another word for *exercise*.

> Exposure to rare words—those not typically a part of everyday conversations—is important for vocabulary development. Children will acquire some understanding of the meanings just by hearing you use them casually in conversations. Helping children build an extensive vocabulary in the preschool years will improve their reading comprehension later on.

Discussion and Shared Writing: Building Signs

- Explain, "Many buildings have signs that give us *clues* about what happens inside them."

- Review the question of the day.

- Show several other pictures of building signs with graphics that provide clues.

- Invite children to read the signs using the graphics for support and guess what happens in the buildings.

- Record their ideas.

- Offer support as needed, e.g., "This sign has a loaf of bread on it. It has two words. 'Yes, Becket Bakery! *Bakery* is the second word on this sign."

English-language learners
Give children time to process language, information, and ideas, and express themselves. This strategy supports all children.

Before transitioning to interest areas, talk about the props in the Dramatic Play area that reflect the inside of the building you visited and how children may use them.

Choice Time

As you interact with children in the interest areas, make time to

- Observe how children use the props in the Dramatic Play area.

- Comment on, or ask questions about, what you see, e.g., "I see you are

setting up a sandwich shop like the one in our neighborhood. What kinds of sandwiches will you sell?"

> **When you talk with children about what they are doing, you make them more aware that they are pretending.**

Read-Aloud

Read *Buildings, Buildings, Buildings*.

- **Before you read**, ask, "What do you remember about the buildings shown in this book?"

- **As you read**, pause to talk about the details in the pictures.

- **After you read**, talk about the buildings that appear at the end of the book. Encourage children to describe what happens inside the buildings in the pictures.

Small Group

Option 1: Bookmaking

- Gather some pictures of interesting buildings.

- Show a picture.

- Ask, "What do you think happens inside this building?"

- If children aren't comfortable guessing, make a silly guess about the building, e.g., "This building has two big circles on the top of it that remind me of a bicycle. So I think it's a factory that makes tires for giant bikes."

- Review Intentional Teaching Card LL04, "Bookmaking." Follow the guidance on the card to have children make a class book about their ideas.

Option 2: Desktop Publishing

- Gather some pictures of interesting buildings.

- Show a picture.

- Ask, "What do you think happens inside this building?"

- If children aren't comfortable guessing, make a silly guess, e.g., "This building has two big circles on the top of it that remind me of a bicycle. So I think it's a factory that makes tires for giant bikes."

- Review Intentional Teaching Card LL02, "Desktop Publishing." Follow the guidance on the card to have children make a class book about their ideas.

Mighty Minutes™

- Use Mighty Minutes 36, "Body Patterns." Follow the guidance on the card, increasing the level of difficulty from the previous day's effort as appropriate.

Large-Group Roundup

- Recall the day's events. Invite children who worked with props in the Dramatic Play area to share what they did today.

Day 5   Investigation 5

What happens inside buildings?

Vocabulary

English: *version*

Spanish: *versión*

Large Group

Opening Routine

- Sing a welcome song and talk about who's here.

Movement: Bounce, Bounce, Bounce

- Use Mighty Minutes 30, "Bounce, Bounce, Bounce." Follow the guidance on the card.

Discussion and Shared Writing: Three Versions of the Same Story

- Show children *The Three Little Pigs*, *The Three Little Javelinas*, and *The True Story of the 3 Little Pigs*.

- Explain, "These books all tell the story of the three little pigs. These books tell three *versions* of the same story."

- Ask, "How are these *versions* the same? How are they different?"

- Record children's responses.

> **Comparing different versions of the same story enables children to use high-level thinking, language, and literacy skills.**

Before transitioning to interest areas, talk about the props in the Dramatic Play area that reflect the inside of the neighborhood building that you visited. Discuss how children may use them.

English-language learners

Children who are not yet speaking English often find dramatic play to be one of the most challenging and stressful classroom activities. During this choice time, you can support English-language learners by coaching children's interactions in the Dramatic Play area.

Choice Time

As you interact with children in the interest areas, make time to

- Observe children as they use the props in the Dramatic Play area.

- Offer ideas and items to extend their play, such as materials to make signs for the building or stickers to use as price tags.

> **Introducing new props is most effective when you select them in relation to a topic in which the children are already interested or already studying. Children are then more likely to work together, sharing common experiences to create their own original play scenes.**

Read-Aloud

Read *The Three Little Pigs*, *The Three Little Javelinas*, or *The True Story of the 3 Little Pigs*.

- **Before you read**, talk about the question of the day. Then read the book that got the most votes from the poll.

- **As you read**, pause to encourage children to retell parts of the story and fill in familiar text.

- **After you read**, read the version that had the second-most votes, if you have time.

Small Group

Option 1: Bookmaking

- Gather some pictures of interesting buildings.

- Show a picture.

- Ask, "What do you think happens inside this building?"

- If children aren't comfortable guessing, make a silly guess, e.g., "This building has two big circles on the top of it that remind me of a bicycle. So I think it's a factory that makes tires for giant bikes."

- Review Intentional Teaching Card LL04, "Bookmaking." Follow the guidance on the card, using the children's ideas to continue making the class book together.

Option 2: Desktop Publishing

- Gather some pictures of interesting buildings.

- Show a picture.

- Ask, "What do you think happens inside this building?"

- If children aren't comfortable guessing, make a silly guess, e.g., "This building has two big circles on the top of it that remind me of a bicycle. So I think it's a factory that makes tires for giant bikes."

- Review Intentional Teaching Card LL02, "Desktop Publishing." Follow the guidance on the card, using the children's ideas to continue making the class book together.

Mighty Minutes™

- Use Mighty Minutes 38, "Spatial Patterns." Follow the guidance on the card.

Large-Group Roundup

- Recall the day's events.

- Share the class book that children created during the past two days.

Further Questions to Investigate

How can we extend the study further?

If the children are still engaged in this study and want to find out more, you might investigate additional questions, such as these:

- What can I find out about my house?

- How are the insides of buildings constructed?

- Why do some buildings have basements or cellars?

- Where do builders get what they need to construct buildings?

- What is involved in painting a building?

- What happens when buildings get old and begin to fall apart?

- How are old buildings different from new buildings?

- What happens at a construction site?

- What are some of the biggest and smallest buildings in the world?

Are there additional questions that will help you extend this study?

Our Investigation

Our Investigation

	Day 1	Day 2	Day 3
Interest Areas			
Question of the Day			
Large Group			
Read-Aloud			
Small Group			
Mighty Minutes™			

Day 4	Day 5	Make Time For...
		Outdoor Experiences
		Family Partnerships
		Wow! Experiences

Our Investigation

Vocabulary

English:

Spanish:

Large Group

Choice Time

Read-Aloud

Small Group

Mighty Minutes™

Large-Group
Roundup

Celebrating Learning

Closing the Study

When the study ends–when most of the children's questions have been answered– it is important to reflect and celebrate. Plan a special way to celebrate their learning and accomplishments. Allow children to assume as much responsibility as possible for planning the activities. Here are some suggestions:

- Have the children work together to plan and build a replica of a building using cardboard. (This structure might be your school building.)

- Open your classroom as a "Building Museum." Have children serve as "docents" and guide visitors through the displays and constructions that they created.

- Take a final field trip to a special building, such as your local city hall or a nearby historical house. Alternatively, you could visit an active construction site.

- Find a carpenter who is willing to come in and help children build small wooden buildings to take home.

- Put together a big class book, photo album, or documentation panel about the buildings study.

- Host a special "Taking Care of Our Building Day." Recruit some volunteers to help small groups of children clean, sweep, wash, scrub, or tidy up various parts of your school or program building.

The following pages provide daily plans for two days of celebration. Add your ideas and the children's ideas for how to best celebrate all of their learning.

Celebrating Learning

	Day 1	Day 2	
Interest Areas	**All:** displays of children's investigations **Computer:** eBook version of *Keep Counting*	**All:** displays of children's investigations **Computer:** eBook version of *Buildings, Buildings, Buildings*	
Question of the Day	What would you like to show our guests about the buildings study at the celebration tomorrow?	Which part of the study did you like best: building lots of buildings in our classroom or learning about our neighborhood buildings?	
Large Group	**Song:** "Dinky Doo" **Discussion and Shared Writing:** Preparing for the Celebration **Materials:** Mighty Minutes 24, "Dinky Doo"; letter cards; list of ideas that the children generated during large-group roundup on day 2 of Investigation 5; "What we want to find out about buildings" chart; Intentional Teaching Card LL26, "Searching the Web"	**Game:** Words in Motion **Discussion and Shared Writing:** Taking Care of Our Building **Materials:** Mighty Minutes 10, "Words in Motion"; list of ideas that the children generated during large-group roundup on day 2 of Investigation 5	
Read-Aloud	*Keep Counting*	*Buildings, Buildings, Buildings*	
Small Group	**Option 1: Writing Poems** Intentional Teaching Card LL27, "Writing Poems"; pictures of buildings; paper and pencils; audio recorder **Option 2: Writing Poems** Intentional Teaching Card LL27, "Writing Poems"; a trip outdoors; paper and pencils; audio recorder	**Option 1: Salsa** Intentional Teaching Card LL36, "Salsa"; (See card for equipment, ingredients, and recipe.) **Option 2: Roll-Ups** Intentional Teaching Card LL37, "Roll-Ups"; (See card for equipment, ingredients, and recipe.)	
Mighty Minutes™	Mighty Minutes 4, "Riddle Dee Dee"	Mighty Minutes 33, "Thumbs Up"; familiar classroom item that is a two- or three-dimensional shape	

Make Time For...

Outdoor Experiences

Physical Fun

- Use Intentional Teaching Card P22, "Follow the Leader." Follow the guidance on the card.

Family Partnerships

- Include families in the buildings celebration.

Wow! Experiences

- Day 2: Buildings celebration

Let's Plan Our Celebration

Vocabulary

English: *celebration*

Spanish: *fiesta o celebración*

Large Group

Opening Routine

- Sing a welcome song and talk about who's here.

Song: "Dinky Doo"

- Use Mighty Minutes 24, "Dinky Doo." Try the letter-card variation on the back.

Discussion and Shared Writing: Preparing for the Celebration

- Talk about tomorrow's celebration. Remind the children that they will be helping to take care of the school building.

- Review the list of ideas that the children generated during the large-group roundup on day 2 of Investigation 5.

- Talk about which jobs the children and guests will be able to help perform.

- Talk about the question of the day. Make a list of children's responses.

- Review the "What we want to find out about buildings" chart from the exploratory investigation.

- Talk about how children may use the Internet and books in the Library and Block areas to research any unanswered questions.

Before transitioning to interest areas, tell children that you will help them gather the items from the list to create displays for family and friends to see at tomorrow's celebration.

Choice Time

As you interact with children in the interest areas, make time to

- Help children gather the items they would like to share at the celebration.

- Assist children with Internet and book searches to help them find answers to their unanswered questions.

> **See Intentional Teaching Card LL26, "Searching the Web,"** for guidance on supporting children with their research.

Read-Aloud

Read *Keep Counting*.

- **Before you read**, ask, "What do you remember about this book? Why do you think it is called *Keep Counting*?"

- **As you read**, ask children to predict which numeral will be on the next page before turning the page. Ask, "Do you see things on this page that we didn't count when we read this book before?"

- **After you read**, ask, "If we keep counting past 10, what else would we add to this neighborhood?" Tell the children that the book will be available on the computer in the Computer area.

Small Group

Option 1: Writing Poems

- Review Intentional Teaching Card LL27, "Writing Poems." Follow the guidance on the card.

- Use the pictures in the buildings display in the Block area to guide children in writing their poems.

Option 2: Writing Poems

- Review Intentional Teaching Card LL27, "Writing Poems." Follow the guidance on the card.

- Take the children outside to look at buildings. Encourage them to use what they see as inspiration for their poems.

Mighty Minutes™

- Use Mighty Minutes 04, "Riddle Dee Dee." Try the counting variation on the back of the card.

Large-Group Roundup

- Recall the day's events.
- Remind the children that they will be taking part in a special celebration tomorrow.

Let's Celebrate

Vocabulary

English: *compare*

Spanish: *comparar*

Large Group

Opening Routine

- Sing a welcome song and talk about who's here.

Game: Words in Motion

- Use Mighty Minutes 10, "Words in Motion." Follow the guidance on the card.

Discussion and Shared Writing: Taking Care of Our Building

- Talk about the question of the day.

- Invite children to share with each other and guests some parts of the study they liked best.

- Explain, "We have learned so much about buildings during our study. To celebrate our learning, we thought it would be a good idea to do something to take care of our building!"

- Reread the list of ideas for taking care of the building generated during large-group roundup on day 2 of Investigation 5.

Before transitioning to interest areas, talk about the displays of children's learning that you've set up around the room. Invite children and families to help take care of the building during choice time.

Choice Time

As you interact with children in the interest areas, make time to

- Encourage the children to explain to the visitors what they've learned about buildings. Have them use the displays as prompts.

- Assist children and families with making repairs to the school building.

Read-Aloud

Read *Buildings, Buildings, Buildings.*

- **Before you read**, ask, "Can you help me read the name of this book?"

- **As you read**, invite children to describe the buildings in the book.

- **After you read**, encourage children to compare the buildings in the book to the ones they explored during the study. Tell the children that the book will be available to them on the computer in the Computer area.

Small Group

Option 1: Salsa

- Review Intentional Teaching Card LL36, "Salsa," and follow the recipe on the card.

Option 2: Roll-Ups

- Review Intentional Teaching Card LL37, "Roll-Ups," and follow the recipe on the card.

> After working on the school building, invite families to stay for an enjoyable snack that the children are preparing.

Mighty Minutes™

- Use Mighty Minutes 33, "Thumbs Up." Try the shape variation on the back of the card.

Large-Group Roundup

- Recall the day's events.

- Review Intentional Teaching Card SE10, "My Turn at the Microphone." Follow the guidance on the card and talk about how much children learned during the study.

Reflecting on the Study

What were the most engaging parts of the study?

Are there other topics that might be worth investigating?

If I were to change any part of the study, it would be:

Other thoughts and ideas I have:

Resources

Background Information for Teachers

We construct buildings to shelter us from the elements and organize and facilitate our activities. A *building* is typically defined as "a structure with a roof and walls that is intended for permanent use." With that definition in mind, we can explore many different types of buildings.

Your knowledge about buildings will help you identify areas to focus on with children. Visualize a few buildings in your community, and consider the following questions:

- What type of foundations do they have (concrete, stone, posts)?

- What materials are used for the exterior walls (wood, brick, steel, vinyl or aluminum siding, stucco, cement, stone)?

- Are there differences in the number, locations, and types of windows and doors?

- What colors are the buildings?

- How many stories, floors, or levels do the buildings have?

- How are the different levels accessed?

- How would you describe the roofs (steeply pitched, flat, domed)?

- What materials cover roofs?

- What shapes are the buildings?

- How old are the buildings? How can you tell?

- Who constructed the buildings?

- Who takes care of the buildings?

- How are the buildings used? How do they communicate their purpose?

- Do any of the buildings' parts appear tailored to a specific use?

Think about the vocabulary used to talk about buildings. While children may not learn and use all of these words, consider introducing them as you talk about buildings.

construction
structure, stability
design, blueprint
foundation, stories (levels)
frame, stud, lumber
stairs, porch, deck, balcony
chimney, smokestack
entrance, exit
joist, beam
floor, ceiling
pipes, wires
window, door, trim
eave, roof, shingles, wall
contractor, carpenter, architect,
plumber, electrician
bricklayer, stone mason, excavator
custodian, engineer, inspector

> **What do you want to research to help you understand this topic?**

Children's Books

In addition to the children's books specifically used in this *Teaching Guide*, you may wish to supplement daily activities and interest areas with some of the listed children's books.

A Year at a Construction Site (Nicholas Harris)

Abuela (Arthur Dorros)

Amazing Buildings (Kate Hayden)

Anno's Counting Book

Apt. 3 (Ezra Jack Keats)

Archabet (Balthazar Korab)

Architecture, Animals (Michael J. Crosbie)

Architecture, Colors (Michael J. Crosbie)

Architecture, Count (Michael J. Crosbie)

Architecture, Shapes (Michael J. Crosbie)

Block City (Robert Louis Stevenson)

Building (Elisha Cooper)

Building (Shelagh McGee)

Building a House (Byron Barton)

Building a House (Ken Robbins)

Building an Igloo (Ulli Steltzer)

Building Heroes (Annie Auerbach)

Building Shapes (Susan Canizares)

Building the New School (Ann M. Martin)

Building With Blocks (Jillian Cutting)

Building With Dad (Carol Nevius)

The Busy Building Book (Sue Tarsky)

C Is for Construction: Big Trucks and Diggers from A to Z (Caterpillar)

Calling the Dove/El canto de las palomas (Juan Felipe Herrera)

The Construction Alphabet Book (Jerry Pallotta)

Construction Countdown (K.C. Olson)

Construction Trucks (Jennifer Dussling)

Construction Workers (Tami Deedrick)

Construction Zone (Tana Hoban)

Hammers, Nails, Planks, and Paint: How a House Is Built (Thomas Campbell Jackson)

Henry Builds a Cabin (D. B. Johnson)

The House in the Meadow (Shutta Crum)

The House in the Night (Susan Marie Swanson)

Houses and Homes (Ann Morris)

How a House Is Built (Gail Gibbons)

Let's Build a House (Mick Manning)

The Little House (Virginia Lee Burton)

My Little Round House (Bolormaa Baasansuren)

The New House (Joyce Maynard)

One Big Building: A Counting Book About Construction (Michael Dahl)

Sod Houses on the Great Plains (Glen Rounds)

Tar Beach (Faith Ringgold)

The Three Little Pigs (Paul Galdone)

The Three Pigs (David Wiesner)

This House Is Made of Mud / Esta casa está hecha de lodo (Ken Buchanan)

This Is Our House (Michael Rosen)

Tonka: Building the Skyscraper (Justine Korman)

What It Feels Like to Be a Building (Forrest Wilson)

Teacher Resources

The teacher resources provide you with additional information and ideas for enhancing and extending the study topic.

Building Structures with Young Children (Ingrid Chalufour, Karen Worth)

Cathedral (David Macaulay)

I Know That Building! (Jane D'Alelio)

Let's Try It Out With Towers and Bridges (Seymour Simon, Nicole Fauteux)

Weekly Planning Form

Week of: _____ Teacher: _____ Study: _____

	Monday	Tuesday	Wednesday	Thursday	Friday
Interest Areas					
Large Group					
Read-Aloud					
Small Group					

Outdoor Experiences:

Family Partnerships:

Wow! Experiences:

©2013 Teaching Strategies, LLC, Bethesda, MD; www.TeachingStrategies.com
Permission is granted to duplicate the material on this page for use in programs implementing *The Creative Curriculum® for Preschool.*

Weekly Planning Form, continued

"To Do" List:

Reflecting on the week:

Individual Child Planning

©2013 Teaching Strategies, LLC, Bethesda, MD; www.TeachingStrategies.com
Permission is granted to duplicate the material on this page for use in programs implementing *The Creative Curriculum® for Preschool*